303 Digital Filmmaking Solutions

303 Digital Filmmaking Solutions

Chuck Gloman

McGraw-Hill
New York Chicago San Francisco Lisbon
London Madrid Mexico City Milan New Delhi
San Juan Seoul Singapore Sydney Toronto

The McGraw-Hill Companies

Library of Congress Cataloging-in-Publication Data

Gloman, Chuck B.
303 digital filmmaking solutions : solve any video shoot or edit problem in 10 minutes or less / Chuck Gloman
p. cm.
Includes bibliographical references and index.
ISBN 0-07-141651-X
1. Digital cinematography. 2. Digital video—Editing—Data processing. 3. Video recordings—Production and direction—Data processing. I. Title: Three hundred three digital filmmaking solutions. II. Title: Three hundred three digital filmmaking solutions. III. Title.
TR860.G58 2003
778—dc21 2003044569

1 2 3 4 5 6 7 8 9 0 DOC/DOC 0 9 8 7 6 5 4 3

ISBN 0-07-141651-X

The sponsoring editor for this book was Steve Chapman and the production supervisor was Sherri Souffrance. It was set in Helvetica by MacAllister Publishing Services, LLC.

Printed and bound by RR Donnelley.

This book is printed on recycled, acid-free paper containing a minimum of 50 percent recycled de-inked fiber.

To my grandmother-in-law (my wife Linda's maternal grandmother), Lauretta Morgan (1908–1998), whose time here on earth made the world a better place in which to live. Since she was my third grandmother she gave love, support, charity, cookies, and tea. She always said that my curly hair was wasted on a boy and allowed me to store my rotting classic Thunderbird in her yard; that deserves at least an everlasting dedication.

Contents

Preface

The world, like digital video, is said to be plagued with problems. For each of these problems, there is a solution. Most of the time we can figure out this "solution," but sometimes we need a little help—that's where this book comes in.

When on a digital video shoot, how do you light a car, motorcycle, human, or animal so they look pleasing? I learned by actually doing and saw what worked and what didn't (most of the time it didn't). I thought if I had a guide to follow that gave me reference on what to do, I could save a lot of time and do it right the first time. I took photos, I made notes, and I compiled a list of 303 solutions to various problems that occur in digital video. With this book, you don't have to guess how to do something. In each instance my way worked (sorry Mr. Sinatra), usually not the first time but after I refined it. Now you can achieve the same results without my loss of hair (unless you don't want it).

The title, *303 Digital Filmmaking Solutions*, is literal—303 different solutions exist for problems you may face in the course of digital production and postproduction. My goal in this book is to help you solve these problems before they happen. Like taking a test, if you walk in knowing the answer (this time *without* cheating), you feel better and more confident. Hopefully, this book will be a confidence builder. Now you can walk into a situation and know what to do. Sure, every solution might not be verbatim for what you may encounter, but I assure you it will be similar. Just tweak any solution slightly and it will meet your needs.

Having been in this business since 1980, a lot has changed and you must roll with the punches or you may miss out. All of the experience I gained throughout the years has taught me that I still have quite a bit to learn. Having shot nearly 800 TV commercials, I still approach each one with excitement, awe, and a little bit of fear. I learn from each one. I may remember something I did 15 years ago, and then apply it to my current situation. When my memory finally fades, I'll keep a copy of this book chained around my neck.

I have used each of these solutions and they have saved me a great deal of aggravation. My philosophy has always been learning by doing. Someone can teach or show you what to do, but unless you experience it yourself, you have not fully grasped the concept.

It's not important if you are a novice, somewhat informed in the world of digital video, or an expert; you never stop learning and growing. Many times throughout this book I discuss working for "the client." Even if you aren't making video your profession, the client is the end recipient or, mainly, the person or persons who foots the bill. Don't let the name "client" dissuade you from thinking some of these solutions may not apply to you. My point is anyone can do any solution in this book whether shooting a video for fun, experience, or for a client. If the word client makes you uncomfortable, substitute the words "my friend," "my family," or simply "myself."

Although not listed in order of magnitude, the solutions in this book are grouped by subject: shooting, lighting, sound, editing, and so on. Feel free to skip around and read the solution that applies to your particular problem, or read all the solutions that may apply to your needs.

The main thing is to use this book as a reference. In compiling these solutions, I learned things on the set that were added to this text. You may never have to deal with nudity or creating special effects on your productions. But if you do, I want you to have a guide where you can find the answers to your questions.

Chapter 1

Preproduction Problems

#1 How to Get Started

Getting started is easy because you're at the beginning. You may not even have an idea or concept, but you want to express yourself by making a video.

The very first questions you should ask yourself when wanting to create a video are: What kind of video should I do? Should I begin with something short like a commercial? A commercial lasts from 30 to 60 seconds and might not take too long to shoot, but it is hard to market once finished. Should I create a short subject? A short is what the name implies: a short video. Do I want to do a how-to, training, educational, or corporate video? Once again, these types have a specific audience and the video needs to be tailored to them. How about a documentary? These are popular ways to begin digital filmmaking and tell a story in the process; TV is full of them. Or should I make something everyone wants to see—a feature? A feature is characterized as a film or video over 90 minutes in length (I won't get into extended dramas or mini-series that aren't considered features). These are long form, involve a lot more, and could be expensive to produce.

Whichever type of video you choose to do, the preproduction is basically the same. What exactly is preproduction? Any work you do before the actual shooting of the video, the preparation, is considered preproduction.

I've worked with a gentleman whose approach to filmmaking was to go out and shoot, and then come back, write a script, and try to sell it to a client. This, by far, is the most difficult way to do a video production. Without having an idea, script, storyboard, or shot list, how can anyone effectively produce a video? Sure, you may get some nice shots, but you aren't making a video.

Some might even ask, don't I need an idea before I can do a video? Yes, you need an idea, but the type of video you want to do comes first. An idea for a commercial is far removed from a feature because you have less time to do a TV spot. Your idea might translate well from one form of video to another, but first decide upon a video format.

Once you've decided upon something, let's say a documentary (the same advice applies to a commercial, feature, or any other type of video), watch what others have done for two reasons: one, to see how it's done, and two, to get ideas. If you want to do a documentary, watch the Discovery Channel, the Learning Channel, the National Geographic Channel, or numerous other channels that showcase documentaries. If commercials are your bag, just watch TV. If corporate, training, how-to, or educational films are more enticing, watch examples of these to learn how they are done. If features are the only thing that interests you, watch hundreds of movies. Whatever the genre, observe what others have done to see what works and what doesn't.

You'll find a lot of garbage out there as well as great work. The elements that make up a video, such as the concept, script, cast, direction, shooting, and editing, all must fall neatly into place or the project fails. Watch the same video numerous times and observe the writing the first time, then the acting, then the direction, and so on. By breaking down a video into these smaller elements, you are better able to see what was effective and what wasn't.

As you'll learn throughout this book, the only way to get good or better at something is to do it. Watching others is important, but you really learn by the hands-on approach. Don't begin a new project without doing your homework first; in that, I mean preparation. This begins at the preproduction stage and goes from there. Hopefully, most of what you'll learn along the way will be common sense, but never cut corners at the beginning or your end product will suffer.

Once you have a type of format in mind, it's time to try to generate an idea. The next solution explains just how to do that.

#2 How to Get Ideas

One of the most difficult parts of creating a video is coming up with new, fresh ideas that will set the world on fire without overtaxing your brain. Television has been around for a long time. A lot of the old standard ideas have already been used, and most of the new ideas out there today are bordering on the bizarre.

The first way to come up with an idea is to be creative and develop your own concept, and the other is to borrow an idea from someone else. For the sake of this book, we will be doing no borrowing. The most sincere form of flattery is imitation, but that's not necessarily so in this business. If you borrow an idea, someone most likely got paid a lot of money for it and probably owns the rights. We will not go this route; it's unfair and if the person is bigger than you are, you might get hurt. Be creative on your own; don't use someone else's ideas.

As I mentioned previously, I recommend that you watch other programs of the genre that interests you to help you come up with ideas. You should watch these not to steal their ideas, but rather to see how others approached the same subject. After watching the programs, ask yourself if you can improve on the concept. Where did their video go wrong . . . or right? Critiquing is important; you can learn from others' successes and mistakes. But remember that opinions are free because everyone has one, and everyone can be a critic. Instead of just criticizing, how can you improve on what you've just seen?

Learn as much as you can about the subject you want to shoot. Read all the literature you can find, check out print ads if they exist, get the product and read the label, visit the person, develop their character, and so on.

If you are doing a commercial, in order to best sell a product, you have to know all there is to know about it. What are its strengths and weaknesses? In a feature, what would the viewer like to happen to the lead? In a documentary, how important is this element to the overall story?

Brainstorming is another way that can help the flow of ideas (a flow is better than a trickle). Sit and stare at something and let your mind wander. What comes to mind? Elaborate on that idea; let it grow. Sometimes it helps to bounce your initial idea off others who aren't as close to the project. How do they feel about it? Does it get their attention? The best way to get ideas is to ask lots of questions—to yourself and others. Answering these will make your script better; that's the next step after an idea.

Sometimes you get ideas when you aren't thinking about how to get ideas. If staring at a movie poster is starting to make you ill, do something else. The idea will still be in the back of your mind. Soon something you see or something someone says will trigger your thought processes again.

Look at the competition's videos again. Whatever they do, try just the opposite. Sometimes this will give you a fresh approach. How do you see the music carrying the video? What facet of the product would you like to see to make you rush out and buy it? Would a kid pitching the storyline from his or her perspective help? Try to see it from the kid's eyes (only taller). Do you want to be told about the background history via an onscreen face, a voiceover narration, a little scenario, or graphics?

It's hard to tell someone, "This is the best way to get ideas; it never fails for me." As soon as you say it works for you, it won't work for them. Coming up with ideas can be done in a thousand ways, and I'd tell you every one of them, if I could. What helps you to think creatively? Go there and do just that.

#3 How to Create a Realistic Budget

The most important question on everyone's mind seems to be how much the project costs. A detailed budget is the only way to determine how much money a video is going to cost to create. The next tough decision is determining what is realistic.

You've heard the old adage that a video production is $1000 to $2000 a finished minute. I've worked on a few projects that were more than that and thousands that were in a far lower range. When asked to shoot any video, you need to determine how much money is needed to complete the project. Too many people may start something but run out of money before completion. A video that isn't finished will never be seen and therefore not make any money.

If you take the time to plan out what everything is going to cost before beginning, you'll have a better handle on whether or not a particular expense is something you should undertake (not undertaker, that's something else). With a detailed budget, you now have a roadmap to follow and can determine where you spend too much or hopefully too little.

Budgeting software programs are available from a number of sources and are laid out like a spreadsheet. You type a number in a particular column and the bottom-line total keeps adding up. However, these programs are expensive for a low-budget digital filmmaker.

Figure 1.1 shows an example of spreadsheet software that is used in the industry. It's a bit expensive, but it is also the best. Some people have saved money and used an Excel spreadsheet. This is essentially how software programs create a budget tool. Like a police officer writing a ticket, leave no area of a budget line blank or it will bite you later.

No step should be ignored in the budgeting process. How do you know how much a production will cost unless every possible expense is listed? Of course, you will have no control over certain variables and expenses, but allowing for contingencies and any possible overbudget items will save you headaches in the long run.

The easiest way to arrive at a budget is to break down the production into three smaller steps: preproduction, production, and postproduction. Let's look at what's involved with each one of these stages.

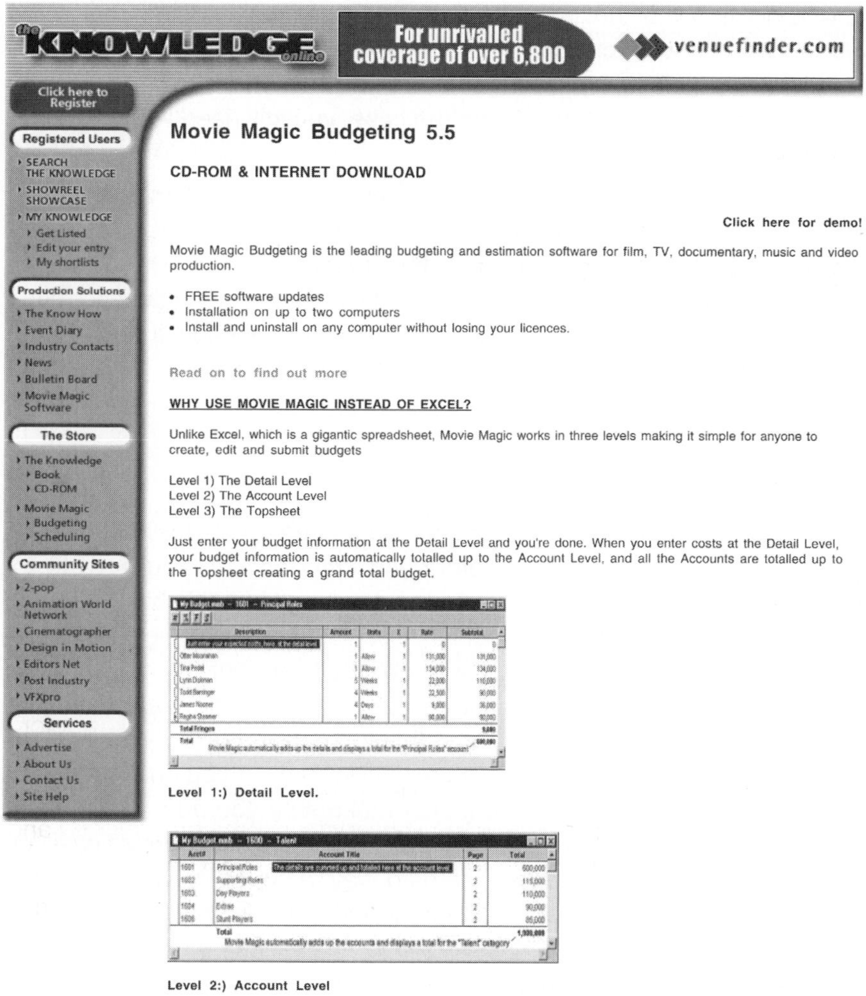

the KNOWLEDGE online

For unrivalled coverage of over 6,800

venuefinder.com

Click here to Register

Registered Users
- SEARCH THE KNOWLEDGE
- SHOWREEL SHOWCASE
- MY KNOWLEDGE
  - Get Listed
  - Edit your entry
  - My shortlists

Production Solutions
- The Know How
- Event Diary
- Industry Contacts
- News
- Bulletin Board
- Movie Magic Software

The Store
- The Knowledge
  - Book
  - CD-ROM
- Movie Magic
  - Budgeting
  - Scheduling

Community Sites
- 2-pop
- Animation World Network
- Cinematographer
- Design in Motion
- Editors Net
- Post Industry
- VFXpro

Services
- Advertise
- About Us
- Contact Us
- Site Help

Movie Magic Budgeting 5.5

CD-ROM & INTERNET DOWNLOAD

Click here for demo!

Movie Magic Budgeting is the leading budgeting and estimation software for film, TV, documentary, music and video production.

- FREE software updates
- Installation on up to two computers
- Install and uninstall on any computer without losing your licences.

Read on to find out more

WHY USE MOVIE MAGIC INSTEAD OF EXCEL?

Unlike Excel, which is a gigantic spreadsheet, Movie Magic works in three levels making it simple for anyone to create, edit and submit budgets

Level 1) The Detail Level
Level 2) The Account Level
Level 3) The Topsheet

Just enter your budget information at the Detail Level and you're done. When you enter costs at the Detail Level, your budget information is automatically totalled up to the Account Level, and all the Accounts are totalled up to the Topsheet creating a grand total budget.

Level 1:) Detail Level.

Level 2:) Account Level

FIGURE 1-1

A budget form from software website www.theknowledge.com/movie.magic.budgeting.asp

#4 Preproduction

Preproduction, as the name implies, is the time spent before the production, or shooting, begins. This stage includes writing a script, storyboarding, auditioning talent, hiring talent, securing locations and sets, building and propping a set, gathering a crew, finding video equipment, and anything else that should be done before you roll tape. Break down each of these substages and decide how much money you can and should put into the budget.

Take the time necessary to sit down and determine what each area will cost. By having the word "script" on a budgeting sheet, it forces you to think about that line item cost. Can you write your own script? If so, should you be paid? How much can you allot out of the budget to hire someone to write it? If the script is handed to you and you don't have to pay for it, you are ahead of the game. If this is not the case, you still need to be aware of it as a potential cost item.

Go through every other item in the preproduction section and decide how much each area is going to cost. Once everything has been added to this section, do a subtotal and see if you can trim costs anywhere or if you need to add money. Luckily, this budget isn't carved in stone because it will change. You'll never be able to know every possible variable that will come into play. The best you can do is head these off at the pass.

#5 Production

Production is where you really start spending money. Instead of one person being involved, you now have actors and crew people. You have to (or should) pay the cast and crew; rent cameras, sound, and lights; use video stock; and cover the location expenses, such as travel, lodging, and meals. This last area is where a lot of money should be spent but normally isn't. On a low-budget production, the least you can do is feed the cast and crew. This can be as simple as snack food, fruit, and having drinks readily available, or on the opposite end, having a catered affair. If you choose the latter, you will be paying per person. Just make sure you add an extra few people to that number. Crew sizes tend to grow and it's embarrassing if you run out of food. But I've learned the hard way; a happy crew is a well-fed crew. It shows you are concerned about your "family" by having food at their disposal.

As each minute of videotape is recorded, the dollar signs are flowing by. The key is to allow yourself enough time to get the shooting done. If you have 20 setups, can you really get them done in a day? If you have budgeted a one-day shoot, and it ends up taking longer, you've lost money. You do need to be realistic. If the locations are separated by a long distance, it does take time to travel from one spot to another. Factors like traffic, weather, and the number of people moving from place to place will also change the amount of time involved. It is impossible to plan and budget for everything, but allowing some play in the final numbers is the safest bet. Budget high—you can always spend less money, but it hurts to ask for more.

#6 Postproduction

Postproduction is anything and everything involved after the shooting is completed: offline and online editing (rough cut and final cut), music, duplication, and labeling. Some people totally destroy the budget in editing. Mistakes in production may be fixed in editing, but usually at a cost. Leave yourself plenty of time in the offline or rough-cut portion of the edit. The hours spent here are cheaper and you can try editing shots in several ways. The online time is too expensive to be changing things at this stage.

Remember to be slightly generous with the amount of time you'll need without gorging the budget. How much time is it going to take to wade through 43 hours of footage? Budget in the cost of your time as well as the use (wear and tear) of the equipment. In this area, I came up with an hourly cost that includes my time as well as depreciation of the equipment:

Operator's Time + Equipment Usage = Offline Cost per Hour

($ ____ per hour) ($ ____ per hour) (Each hour costs $ ____)

Once you have completed your budget, add at least 10 percent to the final figure. This may not sound like a lot of money, but you will exceed your budget in some areas and this is a great place to help fund that overture.

#7 How to Audition Children

When looking for the perfect child for your next video, make sure you audition him or her before you cast the child in the part. If your neighbor, wife, or girlfriend insists you use "their little blossom," make sure he or she can act. Good-looking kids are easy to find; talented ones are more difficult.

Have the child involved in as many aspects of the production as necessary. Tell him or her the purpose of what you're doing (not to make money), what the various roles of the crew are (don't use terms like Grand Poobah or His Worthlessness), and introduce the child to everyone. It's easier for everyone to call the kids by their first names instead of "Hey, kid!"

Some kids have acted before and some haven't. This you usually find out in the auditions. But treat all children the same, professional or nonprofessional. It's exciting to be in front of the camera and make it an enjoyable experience for them, but don't let them draw smiley faces on the lens.

Let the kids wear makeup before being shot (you know what I mean). This makes them feel more grown up. Rehearse the action several times before the cameras roll. This helps put them at ease before you actually begin. In the dry run, do everything exactly like the actual shoot so they know what to expect.

Make sure each child knows who is in control on the set (they are, but don't tell them). Don't talk down to them, but make sure they know exactly what is expected from them. If you relinquish authority early on, you may never get it back again. Try shooting from the child's eye level. That may mean lowering the tripod, shooting from your knees, or digging a small trench. When kids in the viewing audience see other children shot at their eye level, they are more apt to identify with them.

If one of the kids makes a mistake, don't make a big deal out of it. Keep rehearsing until you get the performance you are after. If that still doesn't happen, don't embarrass him or her in front of everyone. Ask the kid if he or she would like a break and discuss the problem privately. If the child is impossible, discretely ask the parents to talk to the child. Kids are kids, but have moods just like adults. Make sure you are with the parents when they explain what the child should do.

Don't work them too long (they'll complain), feed them often (no sugar), and compliment them when they do something correctly (without being condescending). You may be working for them some day and you can always remind them how nice you were to them way back when (they'll forget, but say it anyway).

#8 How to Audition Adults

#8

When your needs call for adults in your video, the auditioning process is where you see what the market has to offer and you get an opportunity to weed out the undesirable talent. In Chapter 3, "Production Software Solutions," I discuss working with nonprofessional actors, but here I talk about auditioning the pros, people who act for a living.

Since you're working with professional adults, the process is much easier than working with amateurs. Contact an agency or organization, or advertise for actors for a particular part. Where this talent comes from isn't important, but what is important is that you select the person best suited for a role.

I find it helpful to have a mass audition where five or more people are trying out for the same role. You may audition for every role in your script at the same time, but screen for each particular role one at a time. When I have a casting call, I usually have several people arrive at a predetermined time. Here's an example of a recent part I had to fill.

I needed a woman to portray a young mother of two for a training video. The role was small, but it still involved acting and spoken dialogue. Seven women from the ages of 20 to 38 arrived ready to read. This is a large range and they could all fit into the role of "young mother." I had a specific idea that 20 was too young and 38 too old. When I walked out and saw all seven waiting, immediately the youngest and oldest struck me as not being right for the part because of their ages. However, you shouldn't discriminate at this early stage. Audition everyone; maybe the people not right for this part could fit into another role at a later date. I've found great talent this way that I might not have normally seen and have added to my casting pool.

When the actors first arrive, give them a copy of the actual script and have them read for that role. Choose a scene that will display the greatest acting ability or emotion in the story. If the part just has them standing there and pointing, your audition process will be that much easier. By having them read and act out a difficult scene, you can determine their range, making the easier parts a breeze.

Call each person onto the set one at a time with you or someone else reading the other roles. If everyone walks in together, some may be intimidated, and some may simply mimic others' auditions.

Always audition in front of a camera. You should do this for three reasons:

1. You see if the camera likes them. I've had people that looked great in real life, but the camera distorted that beauty, while others that were plain looked wonderful on camera.
2. You evaluate how they perform in front of a camera. You probably will be using a camera during the shoot so now is a great time to use it.
3. You now have a record of their audition for future use.

Begin by having them state their name and the role they are auditioning for. This gives you a visual and aural record. Some talent whispers this kind of information and "switches on" their acting ability, while others walk in carrying their acting (or better said, they are already in character). This little bit of information tells you a lot about the person. He or she may be a wimp in real life but can transform into any character. On the other hand, the person may be "on" all the time, always playing the role. Neither of these is better than the other; it just shows what he or she is like as a person.

At times, I prefer actors who are naturals in that they really aren't acting; they are just playing themselves. At other times, I want someone to pretend they are someone else, or to act. Each case is different, so watch for these things.

I always take notes when I watch the live audition, but I will add considerably to them when I view the tape later. This is an evaluation process and no immediate decisions should be made until everyone has completed their auditions. Many times my first instinct about a person was wrong. Upon later viewing I saw something that I had missed in the live audition.

It's also important to have the talent move as much as possible to see how comfortable they are when speaking while wearing a microphone. It may sound silly, but some people cannot walk and talk at the same time.

Some people cannot wear microphones; they feel self-conscious, they are always tugging at it, or they speak directly into it. Make note of these things also.

Try to light the set somewhat so you notice how the characters appear on camera without makeup. Do they squint or are they natural around lights? Will they require a lot of makeup or none at all? Are they comfortable on a set?

Allow them to use the script when auditioning if they need to. It is the first time they've seen it and some are a quicker study than others. Ask yourself questions during their audition: Do they feel comfortable with the character or are they adding or deleting too much? Are they trying to impress you with memorization skills but ad-libbing words that aren't in the script? Are they reaching to play the role in that you can see they are acting?

Most will be on their best behavior, but sometimes you can sense trouble in the making. Are they challenging you with your direction? Will they try anything once before saying no?

Give them guidance on how you see the role being read, and give them some direction. Like cattle, actors need direction. They have no clue what you want them to do unless you tell them. Never tell them, "Just read what is there and I'll tell you if that's what I want." That is not fair to you or them. If you have something specific in mind, tell them and you will also determine at this time if they can take direction. They don't know what you want; only you know that answer.

Without exhausting them, put them through the paces and invite them back for a second audition if their first performance is close to what you were expecting. I once found the perfect person for a role, only to discover I had to shoot everything between the hours of one and three in the afternoon because that's when her child was napping. By finding out their availability and other commitments, you can save yourself a lot of aggravation later.

When the reading is finished, don't expect a polished performance. One thing you will determine is how well they can handle the part with minimal direction. Another great choice (I thought) wouldn't remove her tongue ring for the part. A little give and take should be expected. Tell them what the role entails and ask if they have any problems with that. If their hair is long, would they consider putting it up or possibly cutting it slightly, and so on?

In short, treat them like professionals. After you tell them what you expect from their character, see if they can display that. I've never thrown anyone out, but you will get people who are horrible; that's why you audition. If someone is unsuited for the role, don't embarrass them. Instead, thank them and move on.

I still have never had the perfect audition. It's a learning process for me as well as the actors. Let it be an enjoyable experience for both of you.

#9 How to Write a Video

I believe this is the most difficult question since "How long is a piece of string?" The first step is to determine what are you writing. A commercial is different than a training video, as a corporate production is different than a documentary. Whatever the type of production, your concept or idea should be scripted before any video is shot.

The first place to start is with an idea. Some people can just sit down and begin writing, but even these people have some form of an idea before they begin. If you are doing a TV commercial, start by watching other commercials that feature the same type of product. You should watch these not to steal their ideas, but to see how others approached the same subject. After watching them, can you improve on the concept? Where did their spot go right or wrong? Critiquing is important; you can learn from others' successes and mistakes.

If you are writing a corporate, educational, training, or feature video, the same rules apply. Watch what others have done in the past and see if that sparks any ideas.

If you are doing a corporate video, sit in on meetings to learn more about the project, do research on the Internet, and ask people who are employed at the company. You need to gain as much understanding about the company, product, or place as you can. If you can become an expert, then you can relay that information to the video via your script.

A corporate, training, or educational video is no different in the writing process. It may sound like common sense, but every script must have a beginning to establish the characters, location, and story. The shorter the script, the less time you have to launch this.

The next step is the middle where the story, instruction, or training unfolds. This can be the longest part, but don't get tied up with length. Let the story unfurl and take as much space as you need. If you have a five-minute video and your middle is eight minutes long (you better diet), find out what information may be deleted.

The last step is the ending where everything should be resolved. Upon viewing the finished video, the audience will be educated, trained, or know more about the corporation. Great books are available on the writing process and can go into a great deal of detail on character development, structure, and so on. Read these for learning the basic elements. The important thing is to have all your information presented to the viewer in some fashion, either through narration, dialogue, visuals, or graphics. Which one you use may be decided by your client, but your job is to inform and educate.

Look at the script as an outsider or end viewer would. What would you like to see happen and when? Assume the viewer knows nothing about the subject and you must bring them up to speed quickly. Practice telling the concept of your video to someone before you begin writing. Do they have any questions? What would they want to know more about? Less about?

The best thing you can do in writing any script is to ask a lot of questions. Once these are answered, you will have a script that informs, educates, and entertains the viewer.

#10 How to Storyboard

The most important thing you can do in preparing for a shoot is to create a storyboard. This roadmap is the best tool in your arsenal. Throughout every phase of the production, the storyboard will be referred to, improved, and rearranged. The best way to prepare is to totally have your act together, and the best time to do that is at the beginning. Once you have an idea, it should be storyboarded.

All feature films that are made begin with a concept, then a script, and then a storyboard. With all that goes on and into a feature film, it would be impossible to plan everything out in detail without a storyboard. Whether you are doing a short, a commercial, or a corporate, training, educational, or documentary program, it will require no less. Even though the running time of your epic may be less than 90 minutes, you are still telling a story that needs to be planned out in detail.

Storyboarding should be set up the same way no matter what type of project you are doing. Students in my production classes are always complaining that they hate storyboards, they are useless, and no one else will ever see all their nifty drawings. I also hate doing storyboards, but they are necessary and I would never work without having a storyboard. In the real world, people actually create storyboards (they don't use algebra like I was told they would). They also aren't useless because I refer to mine often, others need the information they contain, and, most important, they are the only common link between everyone on a production.

I can't draw a straight line, but I still do my own storyboards (that's the thrifty man deep inside of me). I can cut out images and I can use stick figures or anything else that depicts the visual element. Storyboards aren't graded on how many different colors are used, how detailed the artwork is, and if it is laminated on poster board (even in my class). I grade my student's storyboards only on the information they contain. Does each frame contain the running time, the action of the character, the focal length, and the transition to the next frame? If it doesn't contain all this pertinent information, I deduct points. Some of the artwork is in color, computer generated, stick figures, charcoal sketches, lipstick smears (from the guys), and so on. I don't add or deduct points for this. Is the storyboard complete, easy to understand, and an accurate portrayal of what the filmmaker wants to convey? If so, then it is a good storyboard.

I wish I could grade some of the storyboards I get from my ad agency clients. They spend a fortunate having these drawn up by an artist and I gain no information from them to help me on the shoot. I show these same storyboards to my students as examples of what professionals try to pull off. Don't worry how pretty they are; just make sure they are complete with all the necessary information.

The length of the production, whether it's a commercial, short, or feature, makes no difference. On a sheet of paper, place the visual instructions on the left and the audio information on the right. Label everything, everywhere, always. If you don't, questions and misunderstandings will arise that will bite you later. On the top of the form, list the pertinent information such as the client's name or account, the working title of the project, the duration, the date, the page number, and any other information necessary to set this particular work apart from the hundreds of other projects you may have.

Some people believe you have to be an artist to create a storyboard. I won't lie and tell you that it *doesn't* look a lot neater and more professional if you have accurate, four-color, life-like drawings depicting the action of the video. But unless you *have to* impress the client with that type of detail, it isn't necessary. What *is* important is that all of the video information is recorded in the video column. Whether it is drawn, typed, printed, computer generated, or in Braille, whatever works for you and the client . . . do it.

I usually type out my storyboards. If I'm really feeling creative, I will cut out pictures and attach them to the storyboard. All camera movements (zoom, pan, or tilt), actor movements (woman walks up and gets into the bus), and prop and set information (one 1964 Ford Falcon, a paper bag of groceries, and a crowbar) are recorded in this column. I prefer to write too much detail in this column rather than not enough. If for some reason I wasn't at the shoot (of course, that would never happen), someone else could look at my storyboard and know exactly what I wanted. If you spell everything out, no one will misunderstand or question it later. Sometimes things look great in a storyboard, but at the shoot it's another story. You have no excuse for poor planning. Come to the shoot prepared with your storyboard and if you come up with something better on the location, at least you have something to use as a guideline.

As a director, I'm not the only person who sees the storyboard. Camera people, editors, continuity people, prop people, set dressers, grips, narrators, actors, and many others all rely on this piece of paper. Other than the script, it has the most value in a production. Even when the shooting is completed, the graphics people and the editor will also need that information. If you do change something at the shoot, get into the habit of making the changes onto the storyboard. The storyboard should be accurate and up-to-date all the way through the production/postproduction process.

The other side of the storyboard contains all the audio information. Type all spoken words, whether they're on- or off-camera, in uppercase. It's much easier to read. In addition, spell out all the words. Don't use abbreviations. Some literal actors will read the abbreviations and pronounce them as such. It's also important to give as many timing cues as possible. If a narrator has to say three sentences in eight seconds, spell that out in the storyboard. If an actor is supposed to read something a certain way, say that in the storyboard. This also gives the talent some guidelines. If you cover your butt at every possible opportunity, you'll never get sunburned down there.

As always, your first draft of the storyboard will probably look nothing like the final copy. But by looking at the storyboard, you can see how the film has evolved from just an idea in the back of your mind to a completed production.

#11 How to Location Scout

One of the most exciting aspects of a production is seeing the location for the first time. Your mind is wide open and you have a positive attitude in that you want to use this location for your project. All the negatives the site may offer aren't foremost in your mind, but you have to acknowledge them anyway.

A location is anywhere you shoot that is not your office or set. One of the ways most people try to save money is by shooting on location. If the location if far from your business, then the costs of travel, lodging, and outside help can make your costs higher. But if your shooting spot is close by, shooting on location rather than using a sound stage or building a set can save money.

One of the first things to decide is where would be the most cost-effective place to shoot. Any area that isn't your place is considered a location.

So how do you find the right location? Is it better just to shoot everything in the controlled environment of a studio, or do you want to deal with the problems that location shooting might bring? Everything must be weighed carefully to determine which best suits your needs.

You know what's involved in shooting at your place of business. You know your assets, your equipment, and your strengths and weaknesses.

There really aren't a lot of problems shooting at your place; you know it well and have used it often. Your place can be where you live, your office, or any area where your stuff resides. Maybe you just want to try some place different or have an idea of where you'd like to shoot. A location can be some exotic port of call, or it can be five feet away from your front door. Now that we know exactly what a location is, how do you find the right one?

Cost usually is a determining factor. Will the budget allow you to travel with equipment, talent, and crew? I recently bid on a Caribbean shoot where I needed a crew of eight (you would be surprised at the number of volunteers I got). It was going to cost two-thirds of the budget just to fly the crew to the location, feed, and house them. Something to look for when scouting is the accessibility of the location to cast, crew, and equipment. It's no fun to lug tons of equipment and people across miles of gritty sand, pebbles, or grass only to find it's now too dark to shoot. Can you get permission to shoot at this location? If you get permission, will it cost anything to shoot there?

Another thing to look for on location is accessibility to power. Can you run AC lines? Do you need a generator, and so on? Decide what you'll need when scouting, not when you're at the shoot.

Preproduction takes more time and can be more important than the actual production. Work out all your possible problems and have backup plans ready. I've been promised the world on a location scout/preplanning trip. When we arrived for the shoot, some of these promised things didn't happen. That's where contingency plans come in. Be prepared for anything. What you don't prepare for . . . that's what will happen.

A location scout shouldn't be rushed. If you only have five minutes to check out a location, you are wasting everyone's time, including your own. How can you possibly check every facet of an ideal spot without spending enough time there? Walk the entire location and look around. Where's the sun? What about sound or visual obstructions, crowd control, access to amenities, and so on? People don't need too much, but what isn't there you may have to bring in.

Once you find the best location, go there during the time of day you plan to shoot. Locations can look very different in the morning and the afternoon. Just like moving to a new neighborhood, you'll have to see what's happening or who's living on the left and right. Will those 40 college kids having a party on the left or the senior citizen high-rise on the right create problems?

Look for clues lying around. Numerous toys strewn about means lots of kids and the noise they may bring, overgrown grass means no one has been present for a while, and workmen and tools mean construction noises. A little detective work and deductive reasoning will go a long way.

#12 How to Get Permits

The rule of thumb is to obtain a permit anytime you're shooting anything anywhere. You also must get permission from the owner to be on the premises, but a location permit is only available from the local police. The word "permit" is part of the word "permission" and that is exactly what you need.

If the budget is low and you're in a hurry, you may think that a permit isn't worth the hassle or expense. Allow me to solve this old wives' tale. The only hassles in getting a location permit, here we go, is simply asking for one. I've gone on hundreds of location shoots and 90 percent of the time I just called the police department, told them what I intended to do (not a bomb threat), and I stopped by later and got my permit.

The other 10 percent of the time, the police may want you to submit a request in writing and that may add as much as an hour to the process. The written requests may be dropped off, faxed, or e-mailed to the station. Either way it doesn't really add any time to the process. In fact, it takes far longer to do almost everything else.

As far as the expense is concerned, in my short career I've never had to pay for a permit. I've even shot in downtown New York City during rush hour on a Friday and the permit still cost me squat. However, hiring numerous teamsters to do other tasks was expensive, but it had nothing to do with the cost of the permit. Obtaining the New York permit did take a little longer, but it is one of the largest cities in the world. It actually took more time to find the correct precinct than to get the actual piece of paper.

Let me clarify one other thing: Do you need a police permit or a location permit? They are both the same in many ways and also very different. Breaking it down, you need to secure permission to shoot in any public or private location, wherever it is. First contact the owner of the business or building and explain your plight. If the city or someone other than the person you contacted owns the property, you need to get a permit from that person or organization. These are examples of location permits.

If your shoot will be disrupting the flow of traffic or you need the massive crowds that will arrive to be held at bay, this is where the police permit comes into play. An easy way to remember what type of permit you need is to ask the following questions:

- Who owns the location? Get a location permit.
- Are the police necessary on the location (you are shooting in a public place, you require crowd control, traffic rerouted, or certain things closed or turned off)? Get a police permit.

Don't be afraid to ask people what kind of permits you are going to need. You may need only one or several, depending on the location and complexity of your shoot. Finding the right office will take you the longest and double-check that you have everyone's permission before you shoot anything.

Many times you may call to obtain a permit and no one at the office knows how to give you one. That does not mean that you no longer need a permit. It just means that possibly no one has videotaped on this site before (or never bothered to ask for a permit). Ask who else you might need to contact. If the owner says you don't need a permit and he or she owns the property, you can go ahead and prepare for shooting. But if you are using the street in front, parking vehicles, and loading or unloading equipment, the police need to be told because no one owns the streets (except Huggy Bear in *Starsky and Hutch*).

Chances are no one will be checking up on you if you are a student shooting in a rural area, but the law states that anyone recording in a public area must have permission or that person is trespassing. Will you be caught if you don't have one? Probably not, but is that a chance you really want to take? Take a few extra minutes and ask who to talk to about obtaining a permit. In the end, you'll be glad you did.

#13 How to Work with a Writer (When You Have To)

Sometimes it's impossible to do everything on a digital video production yourself and you have to know when to call in some help. In this section, I'll discuss working with a scriptwriter when you are too busy, don't know the subject matter well enough, or just prefer to have someone else do it.

I've written numerous scripts where I knew from the beginning that I was in over my head. One that immediately comes to mind is a legal video where the finished project had to steal the hearts of the jury (maybe steal isn't the best word when mentioning a legal video) and prove beyond a shadow of a doubt that this young woman deserved compensation. The script featured a lot of legal and medical terms that I knew nothing about. I could have just written meaningless words and collected my pay, but if any changes were made, I would be lost. I knew that it was time to call for help.

Plenty of excellent scriptwriters can be found out there and the Internet is one of the best places to find them. The site www.mandy.com will be extremely valuable; you just need to type the word "scriptwriter," and names and locations will be displayed. In my case, I was fortunate enough to find a writer locally who knew the legal and medical jargon my video needed.

From the onset, discuss with the hired writer what your role and (in my case) her role would be. In others words, who is going to write what? I would write the meat of the script and she would handle all the medical and legal terms. Break down the script and decide who will do what; most importantly, determine your deadlines. Anytime two or more people are involved in a project, the odds of something not happening are doubled. You know how you work, but when working with another, you may have no idea of his or her work habits.

We decided early on that she would handle the legal and medical jargon in the script and I would take care of the rest of the scripting. When all parties know their roles in the project, less room exists for error.

Also discuss fees before beginning any work. Does the hired gun charge by the page, hour, or word? Do they need money up front or can they be paid when you are? As long as they are on the clock, you and your budget will be charged; know all this information ahead of time.

Another big snafu with written scripts is changes. If you turn in a script and the client wants a rewrite, how much is that going to cost you? I've worked on projects where the client didn't know what they wanted and I delivered five revisions of the script before it was accepted. If the client is being nitpicky, who should foot the bill for that? Discuss this with your writing partner and see how a scenario like this would play out.

The important thing is to know a ballpark cost figure before you begin. If you budget $1000 and after rewrites you owe the writer $5000, something was lacking in the communication process. Some writers want less money up front as their fee but want a percentage of the profits once the video is completed. I never feel comfortable with this because it could mean a lot of money (coming out of your profit margin) and a much longer relationship with the writer.

You also need to leave yourself an out. What happens if you and the writer don't get along? Anytime two or more people are working together, personalities come into play. I hired a writer that did great work, but he only worked or could be reached after midnight. Having a schedule that was opposite of mine, it was very difficult getting anything done because we were on different poles. "I write better when I smoke, drink, am with people, or am asleep." Everyone has their quirks and I think I met them all.

Contact information is critical because sometimes changes need to be made immediately to a script. Another of my hired guns decided to take a week off and go to the beach the week of our deadline. He was lounging on the beach while I was trying to put out a major fire. It seems I always learn the hard way first, and after I've been burned (from that fire), I rarely get burned again. Have a cell phone number (the most immediate form of contact) or e-mail address (a slower form if they don't check their messages often), or just show up at their favorite hangout.

What happens if the project folds before completion? Do they get paid for what they've done (they should if you have)? These possible contingencies also need to be addressed by all parties. You can't come up with every "what if" because every project is different.

When you begin the writing process, determine a schedule you both agree to. Should you meet and write together? Should you send material to each other as you finish it while working alone? It doesn't matter how it gets done as long as you both agree on how that should happen. Sometimes brainstorming or bouncing ideas off each other creates the best concept. Do what works best for each of you.

My method of writing is usual. When I begin a script or story, I have no idea how it is going to end until I get to that point. I drove my instructors nuts in college because the "correct" way is to do a treatment or synopsis and develop the script from there. When I work with a real writer, they usually prefer to work that way, so I change my habits. I can write the natural way if I need to, but I prefer my own approach. Once again, this give and take is important.

Lastly, you now have two styles of writing, two egos, two salaries, and two roles in this scripting process. The key is to establish milestones and meet them on time and on budget. The rest of the process is easy. Talk to your partner in crime and the only problem you will have is the one that wasn't discussed.

I don't care how writers get the job done as long as that's what happens. Whether they write in the bathtub, while they dream, or while listening to a boring lecture is immaterial as long as I get my material. It's almost impossible to find someone with the same writing habits as yours, so respect theirs.

#14 How to Use Production Software

Although it's not really cheating, production software enables you to complete your given task in less time than the folks doing it the old-fashioned way—by hand. Various types of software are out there, and each one excels in its own area.

#15 Preproduction

These programs help you from the onset of your project. Screenplay Systems, Inc. (www.screenplay.com) offers a multitude of programs to ease you into your video production. Movie Magic Budgeting is a fantastic program that is "the world's leading film and television estimating tool." The budgeting information I mentioned earlier in this chapter is spelled out for you in a database that includes everything you'll ever need to budget.

The same company offers Movie Magic Scheduling with included breakdown sheets, call boards, call sheets, and shooting schedules for costumes, the cast, props, sets, stunts, and vehicles needed for every scene. Although these program cost a few dollars, if you do any number of projects a year, these will help take the tedium out of this extremely necessary and vital step.

#16 Production

Once the "pre" work has been completed, additional software programs can ease your burden in this area. Screenplay Systems, Inc. leads the way again with Movie Magic Screenwriter. This has got to be the easiest way to format scripts without having to space and tab everything. Other programs out there can do the same thing, but this particular package auto-corrects typos, creates character names, and auto-recognizes lists of scene headings, the time of day, locations, transitions, and extensions without having to type any letters. Whether shooting a feature or a 30-second TV spot, the computer makes your output faster, more convenient, and organized.

I reviewed video-logging software called TPEX from Imagine Products, Inc. All video footage shot on location or in the studio is played from your camera directly into your computer, automatically logging the in and out point of every shot as well as generating a thumbnail image icon for each shot. Remember when someone had to watch all the footage and manually write the in and out points of each shot by hand?

As recently as last year, many of these programs were in their infancy and required so much time to learn that you could not save any time using them. That has changed; now the learning curve is almost immediate and you can begin saving time as soon as you open and install the software package.

#17 Postproduction

Software in this arena has been around for quite some time in the guise of *nonlinear editing systems* (NLEs). It's very common to take a laptop computer to a shoot and have a rough cut of the video completed by the end of the session. The platforms of PC and Macintosh are no longer that different and excellent software applications are available for both. Final Cut Pro is the champion NLE package from Apple with Adobe Premiere and AVID DV-Xpress working in both platforms.

I really can't see a downside to using any of the production software available (except possibly the cost). It will take less time when using software, giving you more time to devote to other aspects of your production.

#18 How to Copyright Your Work

Besides hiring someone named "Knuckles," copyrighting your work is the best protection you can get. Once you have a copyright, you are legally registered in Washington, and anyone who tries to use your work other than through you will be in trouble.

What does it take to copyright your work? Very little actually, but you must decide a few things. Do you want to copyright your script or the completed video? You may copyright just one or both, and the process is the same. Actually, the first step you take in the copyright process immediately begins protecting you, and it doesn't cost a cent.

If you decide to copyright your script, the first step is to put the copyright symbol (©) on the title page of the script with the copyrighter's name afterwards. An example would be ©2004 Wonder Pictures, Inc. Just putting this information on your script means that the work is under copyright and is on file in the Copyright Office in Washington, D.C. Although you may not have yet filed a copy with them, having the symbol on the paper means the first step has been taken and the unsuspecting public has no idea if the symbol is all you've done, or that your work is actually registered with the Office. Either way it means "hands off."

The same process applies if you are copyrighting your video. Preferably on the title screen and also at the end of the credits, place the same symbol and information. Figure 1-2 illustrates a sample video title screen with copyright information.

Figure 1-2

Title screen with copyright (for those with good eye sight)

Now that the first step has been completed, step two will cost you a few dollars. Your work needs to be registered at the Library of Congress and they need a copy for their files. In other words, contact the Copyright Office (www.loc.gov/copyright) and they will send you, free of charge, a copyright application. Fill out all the information and decide if you want to register just the script or the entire video. Obviously, the whole video will offer more security from copyright infringement, and if someone wants to make a book from your screenplay, it will be easier than trying to produce another video.

When the form has been filled out, send a copy of what you want to copyright, along with the $25 fee (as of this writing) to the Copyright Office. That's all there is to it. The magical © will be on your work, telling the world that it has a copyright, and a copy will be filed in Washington in case anyone wants to make sure of that fact. Either way, you now hold the copyright for a long, long time.

#19 How to Schedule and Keep Track of Your Production Requirements

Without a detailed schedule, you are dead. With a video production waiting in the wings, everything hinges on you scheduling the right thing to happen at the correct time.

Whether you use a software program mentioned earlier in this chapter or do it by hand, you must keep a running list of when things are going to happen. A feature film would never get made because of all the scheduling logistics unless someone kept detailed records. Your digital production is no less important and also needs detailed scheduling.

Like building a house, you must schedule certain things to happen before others can occur. Without the framing being completed and signed off, the drywall can't be hung. The painters can't paint the drywall unless it's hung and so on. Actors must be auditioned and hired before shooting can begin. How many days of shooting are needed at each location? When is your star needed on the set and when can you schedule "lesser" people? Do you really need the Steadicam for three days or would two better suit the budget? How many crew members do you need on a particular set at a given time?

Everything, no matter how unimportant, must be scheduled to determine who needs to be where, when, and for how much. Creating a budget is a lot like scheduling. Every line item you need to purchase must be listed and everything that is going to happen by anyone must be on the schedule.

If I'm the director, I need to know when I must direct which scene. After each item on the schedule occurs, that must be noted. Take a close look at the great budget you developed. The people in that budget should also be on the schedule with what they are expected to do and when. Which scene is to be shot at location five? How many actors are needed in Scene 3b?

Like a daily planner for anyone in business, your production schedule is the only way you or anyone is going to know what's happening.

Chapter 2

Production Hardware Solutions

#20 Production Hardware Solutions

Finished with the preproduction phase and not quite ready for production, you are going to have to address problems with hardware, the equipment you will be using on your shoot.

If your shooting involves an image that moves closer to and farther away from the camera, what can be done? Or if the object just sits there, how do you add a little onscreen life to it? This chapter will also discuss which type of lens to use and when, how to use a moving camera to pull the viewer into the shot, and what a Pro-Mist filter does to the look of an image. It will also look at how and when a particular type of filter should be utilized, how to transform one type of lighting instrument into another, when to use glass as a diffusion material, when to use *halogen metal iodide* (HMIs) on a shoot, and which HMI is best for a particular situation. The chapter also examines using multiple cameras on a shoot, being aware of what each one is doing at all times, using video assist when shooting film, and how to keep sand from getting into areas it shouldn't.

#21 How to Solve Production Problems by Knowing Your Equipment

Once your preproduction work is completed (you have a script, storyboard, actors, crew, location, and equipment), you are ready to begin production. This is also the time that a whole new set of problems will arise.

The role of a producer on a production is someone who must solve problems. This chapter will discuss various production-related problems and how to solve them without calling a halt to the shoot. Most of these problems are nothing a producer can solve. They are equipment related and should be someone's responsibility on the set. Once you are on location (a studio or someplace else), the cost for talent, crew, and equipment is escalating. With this in the back of your mind, you must calmly solve the problem at hand without worrying about how much the delay is costing the bottom line.

Although none of the problems I will be mentioning in this chapter are that complicated or earth shattering, they are hardware related, and *before* the shoot actually begins is when these should be addressed, not when you're ready to start. It's crucial to be familiar with your equipment well in advance of the shoot so time isn't wasted on the set.

On a recent corporate shoot, I had to videotape the installation of a laminate floor. As the installer would lay down pieces of flooring, I had to shoot close-ups of him applying a bead of glue to the edge of the board's length. This meant that my depth of field would be extremely narrow and I would have to roll the focus from one extreme to another. I knew this was coming because I saw it in the storyboard.

Instead of waiting until the day of the shoot, I practiced beforehand on how to smoothly pull off this feat. I didn't have the installer or the piece of flooring, but I obtained something of equal length and had someone run their finger along the edge of the board at various speeds to simulate the glue applicator tip. My task was to keep his finger in focus if he went forward or backward.

By practicing, I learned which way to roll the focus ring when his finger moved farther away, and the opposite when he moved closer to the camera. This first step is knowing the direction of focus ring rotation; that will never change no matter what you shoot. The other obstacle is the speed. This is something you must anticipate, but with practice it almost becomes second nature.

On the day of the shoot I was able to perform this maneuver more easily because I was more familiar with the camera. Once again, this is where preparation (reading the storyboard, knowing what's involved in the shoot, and becoming acquainted with the lens movement), and preplanning (I did this before the actual shoot) saved time and money on the set.

Obviously, you cannot anticipate every problem you many encounter, but if you do your homework (not in homeroom) the problems may be kept to a minimum. When other problems crop up, don't panic or lose your head. Just break off a small piece at a time and solve it.

#22 How to Maintain Focus on Constantly Changing Planes

One of the problems with shooting is that most people prefer to view the image in focus. But as people move, the focus changes (I love my job so I am always focused). Instead of relying on your camera's auto-focus, how do you keep your mobile actors in sharp focus as they wander the set?

During the course of the day on an outdoor shoot, your f-stop will change. On a recent outdoor shoot, by the middle of the afternoon our available light only gave us an F3.5 when shooting medium shots of our talent. With anything longer than a 50mm lens (35mm film), the focus had to be racked.

A great selling point about film versus video is that on film you are focusing on ground glass (what you are seeing is what you get) instead of a tiny video monitor. Sometimes it's nearly impossible to get sharp focus outdoors while viewing a video eyepiece. On the other hand, once you stop down a film camera's lens, the eyepiece also darkens. When focus is critical, view the image on as bright a monitor as possible. If you have the luxury of follow focus, a device that allows someone to control the focus as the talent moves, make sure the operator has a detailed image in which to view the subject.

In order to get the most accurate focal distance in film or video, you should use a tape measure. Professionals (and me too) always check the distance with a tape measure because the small marking on the barrel of the lens may not always be accurate. Run the tape from the subject you want to focus on to the film plane or video camera's *charged coupled device* (CCD). This is always found a few inches behind the lens.

If your talent is moving, the measured distance will also change. This is where several measurements should be taken at various points. When the actor hits a particular mark, the lens should also be at that focal point. Try shooting with the highest possible f-stop (above 5.6) and with the shortest focal length lens to give you a greater depth of field. The farther something moves away from the lens, the more you should rotate the focus barrel clockwise. As an object approaches the camera, rotate the focus counterclockwise. This concept will never change with any camera or format. If you practice this and commit it to memory, follow focus will be as easy as riding a bike in your wet sneakers with long pants.

The obvious key to follow focus is the speed at which it happens. You must precisely match the speed of the actor (or their speed match yours) to what the camera focus is doing. This does take practice and should not be attempted on the fly.

On all my shoots, my assistant cameraman Greg Ressetar methodically checks the actual distance between the camera and actor with a tape measure and compares it to the lens setting. Since my eyesight isn't what it used to be, if there was any variation in my focal length and the tape measure's, I usually lost. Most of the time, however, I wasn't more than an inch off (I'm not hopeless yet). This technique also works with hand grenades.

#23 How to Create On-Camera Movement with Inanimate Objects

How do you get a wagon, laden with jars of money, to roll down an embankment (on cue), trailing money, and turn off-camera at the precise moment? Of course, there is no guaranteed way, but you can come darn close and increase your knowledge of fishing at the same time.

Place the camera on a high hat to magnify the slope of your incline. This adds to the excitement and makes things appear to be moving faster than they really are. Make sure the path is free from extraneous debris that might alter the wagon's course. Our wagon was laden with phony money and carted to the top of the hill. In the script, the wagon rolls away on its own and starts careening down the hill. A 60-pound fishing line (complete with fishing pole) was used to guide the wagon and keep it on course (the straight and narrow).

In the first rehearsal, the wagon was released and traveled eight feet and disappeared in the woods. On subsequent takes, the steering mechanism was locked so the wagon would travel in a straight line, instead of turning at the wrong time. Trying to catch something spontaneous, the camera rolled as the wagon was released. It traveled directly toward the camera, hit a bump, turned away, and then came back into the path of the camera as it bumped along, dropping bills.

No perfect way exists for moving an object like this down a leaf-covered, unleveled embankment. The fishing line will help you control the direction of the descent, and locking off the wheels will keep it moving in a somewhat straight line. But the best bet is to try numerous takes, recording all of them. Eventually, you will get a better trajectory than you initially imagined. Just protect the camera because that's where a moving object always wants to travel.

Fishing line, because it's transparent, can be used whenever you want to "hide" the wires. If we had the time, we would have devised two lines to run parallel to the wagon, creating a guardrail track on its left and right. The wagon would bounce off the taut line and remain on course as it descended the incline.

Figure 2-1

A wagon in the woods with a string

Use whatever you can to guide the path of a moving object. Remember it will always take the path of least resistance, so give it very little to chose from. In Figure 2-1, the little red wagon prepares for its journey on a string.

#24 How to Make a Lens Work for You

Because a lens enables you to see the world from any perspective, why not let it work for you? You can use a lens to flatten or distort as well as capture a normal image. You know the basics of what lenses can do, but how can you make them do something unique?

You've all seen close-up images shot with a wide-angle lens. It's like looking at someone through the little eyehole on your front door. This is also a wide-angle or wide-field-of-view lens. Features are exaggerated as the camera or person moves closer. Long noses seem to bend around the frame like a funhouse-mirror image and you see a lot more of what you might not want to. A wide-angle lens, when moved in close, is great for humor.

For a humorous commercial, we needed to shoot a little girl (isn't it nice when you say "shoot" in this context?) in close-up but also have her features distort. Her impish, toothless grin, long bangs, and freckles were the "sight gag" the casting director had succeeded in finding. In the client's storyboards, they wanted the child's head to be large and distorted with her feet appearing as a small point. This can only be accomplished with a wide-angle

Figure 2-2

Large head, small body because she is so close to the lens

lens. Keeping her head large in the frame meant having her look up into a fully extended camera. Figure 2-2 illustrates how this was done.

Use as wide a lens as possible without resorting to a fisheye (down to 10mm in 35mm film or 6mm in video). When we used a 14mm lens, we could see her entire body in my frame, but also my feet, her father's feet, the tripod, and the edge of the lake. Since most wide-angle lenses focus as close as 10 inches, move the camera that close to your subject. I know it seems awkward and you are violating her personal space, but do what it takes to get the shot.

With a film camera, shooting this close presented a problem in camera noise. But by covering the camera with numerous sound blankets, the noise disappears (as you slowly bake to death).

A wide-angle lens tends to accelerate motion. Car races are shot with a wide lens to make the cars appear to be traveling at a higher rate of speed. Shooting wide from below gives the illusion of power and dominance (the good guy shot from below, tall buildings looking more ominous, and so on).

The opposite is true with a telephoto lens. Any long lens will compress the image. Remember the famous shot in *The Graduate* where Dustin Hoffman is running toward the camera but seemingly gets nowhere? With a telephoto lens, the actor is traveling a great distance, but because of the magnification and compression, it looks as if little distance is covered. Traffic jams are always shot in telephoto because it makes it seem more congested. The magnification allows rising heat (and exhaust fumes) to be seen.

Anyone with large features (okay, a big nose too) can be made to look more flattering with a telephoto lens. These objects are compressed and look more pleasing.

The next time you are trying to create a visual image, try using a wide-angle or telephoto lens and move the camera accordingly. You will get the desired effect you're after.

#25 How to Use a Moving Camera Effectively

A moving camera doesn't just add fluidity to a shot; the whole production value of the video is enhanced. One of the most underused pieces of film or video equipment has to be the jib arm. Everyone and their brother have used a dolly at some time. When our budgets are really low, we often use a wheelchair (ask the occupant nicely first). They're inexpensive and are relatively smooth, but you still have to handhold the shot, which adds shake. If the budget allows the extra $25, we rent a doorway dolly.

The Matthews unit we usually use also serves the purpose, but if you want to make any turns, you have to decide a week ahead of time. Eighteen-wheelers have a smaller turning radius. And after the camera, tripod, teleprompter, and person climb aboard the dolly, only a steam engine can move it.

But the cream of the crop in our small rental world is the panther dolly. This unit can be used with pneumatic tires or dolly track. The turning radius is like a large General Motors vehicle; it can be pushed, pulled, or slid, and it has a built-in pedestal that accommodates my favorite attachment.

If the client requests or has enough money for a dolly, the next step is to try to get the jib arm. The jib allows lateral, horizontal, and vertical movement. Many times on our shoots we rarely use the dolly in the shot's movement; the jib arm allows us to get closer and usually makes the movement more effective. As long as you balance the jib arm for the weight of the camera, you're in for some pretty amazing shots.

In one commercial we did, every shot in this 30-second opus had to be moving; it didn't matter how slight the movement, just as long as movement existed. I usually prefer using track with the dolly instead of the pneumatic tires. The tires always seem to find every bump in the floor. The track, once shimmed and tightened, is as smooth as glass (or silk if you can afford it). Once the track is in the right place, the dolly is added and the jib arm attached.

With careful balancing, a slight push or pull movement on the jib gave us a smooth shot. We now could swoop down or jib up to get the shot. If you really want the client to rent a jib arm, put the words jib shot, jib up, or jib down in your storyboard.

It's my own fault, but I thought it would be easier to move the 16-foot expanse of track if we folded (or collapsed) it. I picked up one end (by the inside) and the cameraman picked up the other end 16 feet away. Before I could finish saying, "Let's fold it," he already snapped his end closed with my fingers still in my end. I saw colors I had never seen before and my fingers are really popular at pancake breakfasts.

You don't have to ride the dolly, although many think it's cool to do so. The jib can be controlled from the camera position, the back end by the weights, or by sitting on the dolly. I find it easier to watch the monitor in a jib move, rather than trying to keep my eye attached to the camera. Practice the many different methods and find the one you like best. It's almost like using a Steadicam; you'll find your own niche and you'll get better at it each time.

Don't try to walk through the track (between the cross-ties), push the dolly up hill, or push it by the weights. Our intern, "Thumbless," learned this the hard way. The best place for pushing the dolly is the pushing bar, or if you're using the jib, walk outside of the track and push, pull, or steer with the jib itself. Practice until you find a method that works for you.

Clients will also fall in love with this contraption. Once they try moving it themselves to "really see what the shot looks like," they will want to use it on every shot.

In the commercial we did, without the use of motion control, it was nearly impossible to pull or push the dolly slow enough to get a movement without jerks, stops, or starts. No matter how many times we tried, we just couldn't get the movement smooth enough. As anyone knows, hovering is smooth, not bumpy.

The client suggested we wedge a Matthews C-Stand arm between the dolly's pulling bar and the dolly track itself. With a slight push or pull, we now had a fulcrum in which to pivot the dolly. With a slow lifting action, we could push the dolly a total distance of four inches, just enough for the length of the shot.

#26 How to Best Use a Pro-Mist Filter

People are never happy with what they have. You have a great digital camera, but you want a softer, nonvideo look to the piece. One of the best ways to achieve something like this is with a Pro-Mist filter.

In order to soften the hard edges of a digital camera's image, a Tiffen Full Pro-Mist filter may be added. Digital video has a different look than analog. A slightly cooler flesh tone and sharper edges give digital a more newsy reality. In the old days, a tube camera and analog video would have been the only way to achieve this warm and fuzzy look.

Since digital is commonplace and has many advantages over analog, I firmly believe that when shooting digital video, a Pro-Mist is the best filter for softening some of the video harshness or "edge." I will never use anything else in front of the camera. If a full Pro-Mist is too much, try using a half or quarter. This look is a subjective thing, but you will be astounded by the results.

A Pro-Mist doesn't make a digital image look just like analog. If you wanted that sort of look, shoot in analog. Digital is sharper, cooler, and more defined in its acquisition, and the filter slightly softens the edges. Like a cool drink on a hot summer day, this filter takes the edge off but still leaves a high-quality image.

Today we have digital equipment that can give us that analog look without the cost and weight. It's sad to say it took us this long to come up with a look we used to have. Figure 2-3 shows an image without a Pro-Mist filter and Figure 2-4 is with the filter in place.

Figure 2-3

An unfiltered image

Figure 2-4

The same image with a Pro-Mist Filter

#27 How to Use Filters

I've always been an advocate for shooting without filters in the field and changing their appearance in post. The footage may look better if you use filtration on the shoot, but if for any reason you decide to remove the effect, it's next to impossible to do so.

Sometimes you know you want a specific look and no power on earth will change that. This might be the time to use filters. Some of my shooting friends insist on a Pro-Mist filter whenever they shoot digital video because it takes the edge from the crisp CCD look. When I'm on a 35mm film shoot, the director I work with also demands a Pro-Mist to degrade the 10 megapixel look.

The most popular filters are found in three categories: color correction, diffusion, and special effect. Any filter ever made fits into one of these three areas. Let's look at color correction first.

Anytime you are changing the look of your video through color manipulation, that's color correction. These filters, usually numbered, range from the blue series (80A, B, and C) that converts tungsten to daylight, the orange filters (85, coral, and so on) that change daylight to tungsten, and a myriad of others that correct color (minus green, fluorescent, and so on).

Diffusion filters take the hard edge off the video image with the likes of fog, low contrast, mist, and neutral density. These filters each do the job that is implied in their names and the higher the number (Fog 1, 2, 3, 4), the heavier or stronger the effect.

The special effect filters are a little more exotic, such as star filters, gradated, soft focus, distorting, and split field. Color filters like yellow, red, and green produce special looks to your shots rather than color correction (unless you live on a different planet).

Each of these filters will do exactly what it is called upon to do and nothing else. Most of these looks can also be added in post, allowing even greater control of changing the shot. As always, the key is moderation. If someone notices you've used a specific filter, your secret is out and the magic has been lost.

#28 How to Transform an Open-Faced Light into a Fresnel

Sometimes you don't have the type of light you really need for a shoot. With a little ingenuity, you can create the new light from your present arsenal.

Our lighting pooh-bah, George Winchell, created a Fresnel-type light out of a Lowel DP. By attaching a piece of 216 diffusion to the DP's barn doors with clothespins and wrapping the opening in black wrap, George invented a soft, controllable, focusable Fresnel light. Although not particularly attractive looking, these black-wrapped quasi-Fresnels cast a soft glow on the talent. Figure 2-5 shows our newly created invention.

As you know, Fresnels are great for the kind of look their light has. The glass lens on a real unit can create a texture on the talent's face that's hard to duplicate any other way. The diffusion on our open-faced light also creates a texture, but a different one than a glass lens. If diffusion other than 216 is used, the look will also change. Having various diffusion materials in your kit will allow you to devise the look you're after for a particular shoot.

#29 How to Use Glass as a Diffuser

If you have a slightly larger budget but a much larger and darker area to light, you need stronger lights. I don't own stock in HMIs, but sometimes they are the only units that fit the bill.

Once during a particular shot, the talent had to stand on a balcony two stories up and required illumination. The face of the building and the talent

Figure 2-5

An open-faced Fresnel

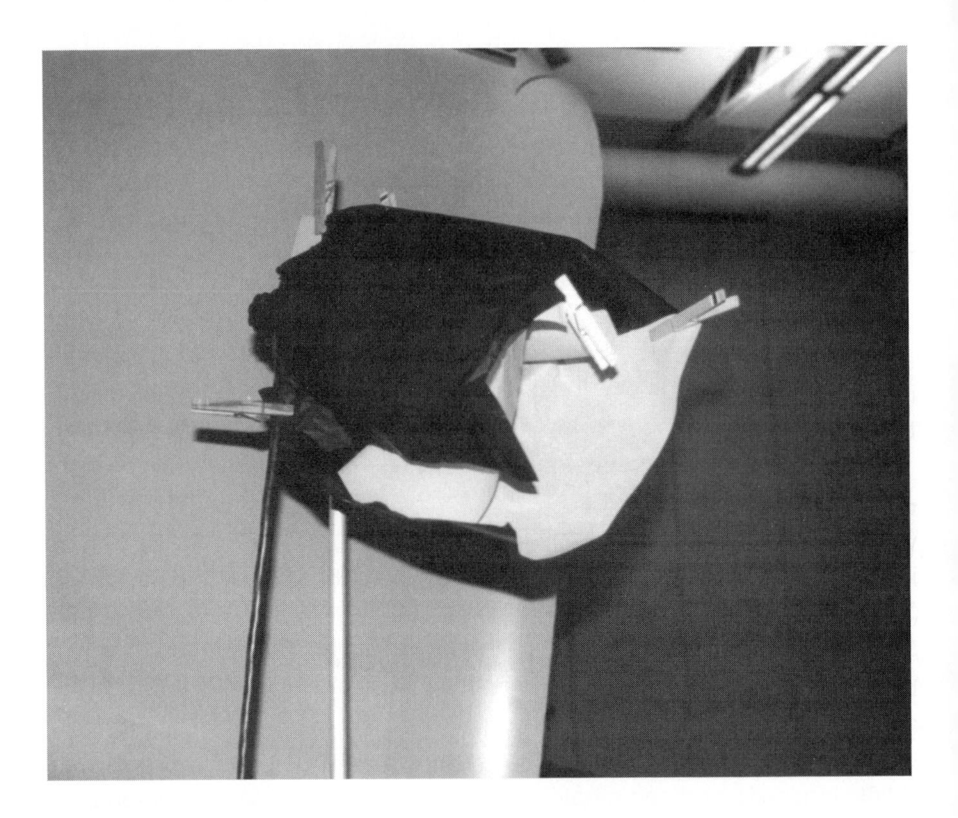

were in complete shade, and the nonbalconied side was bathed in brilliant sunshine. Since it was cheaper to rent HMIs than to move the balcony, we opted for the HMIs. Our DPs wouldn't have enough punch at 1000 watts. The output of a 1200-watt PAR is much brighter and whiter than a DP. At a distance of 6 feet from the talent and 25 feet above the ground, the HMI had to be heavily diffused (and placed on a platform).

HMIs are daylight balanced and much more powerful than conventional tungsten lighting. Our rented Arri 1200-watt PARs were almost too much light. The thickest glass screen was positioned in front of the light, and a folded piece of 216 was clothespinned to the barn doors. Several glass screens come with HMI PARs so experiment with the look you like best.

Another HMI 1200-watt PAR was used as a backlight. This unit was raised 10 feet and positioned 12 feet behind the talent. The same glass and 216 diffusion were also used on this light.

Because of the daylight balance, no blue gels were needed (which would also suck up needed light). In my opinion, the best choice for outdoor lighting is HMIs. They need no color correction, are bright enough (no dumb lights allowed), and have built-in diffusion with glass screens.

#30 How to Use HMIs Effectively

There's really nothing negative you can say about HMIs. I've been using them since I started in this business and sometimes they are the only lights that will do the job. Although Arriflex manufactures plenty of different sizes, models, and types of HMIs, I'll only be discussing two types here: the PAR and the Fresnel.

The one thing that separates HMIs from the rest of the pack is that they are daylight balanced. Once an HMI fires up, you'll notice that the light is much whiter to the eye than a 3200K unit. As I've mentioned numerous times before, in video, whites really pop and fluoresce when using an HMI.

Whenever you need an extremely powerful source of light that packs a punch, when you have a majority of daylight on your location, or you just want a purer light that does the work of many others, the only place to look is the HMI. Once you have used it on a shoot, you will be as sold as I am on their capabilities.

The only difference in setting up one of these units is that the light must be plugged into a ballast. Once warmed up to the correct color temperature and intensity, these brutes will provide pure light all day. But how do you use them effectively? Figures 2-6 and 2-7 show an Arri Fresnel and PAR respectively.

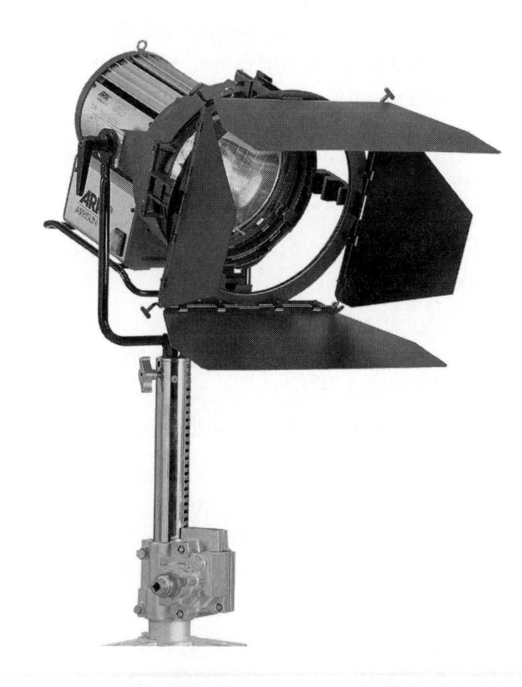

Figure 2-6

An Arrisun 12SB R HMI (Image Courtesy of Arri, Inc.)

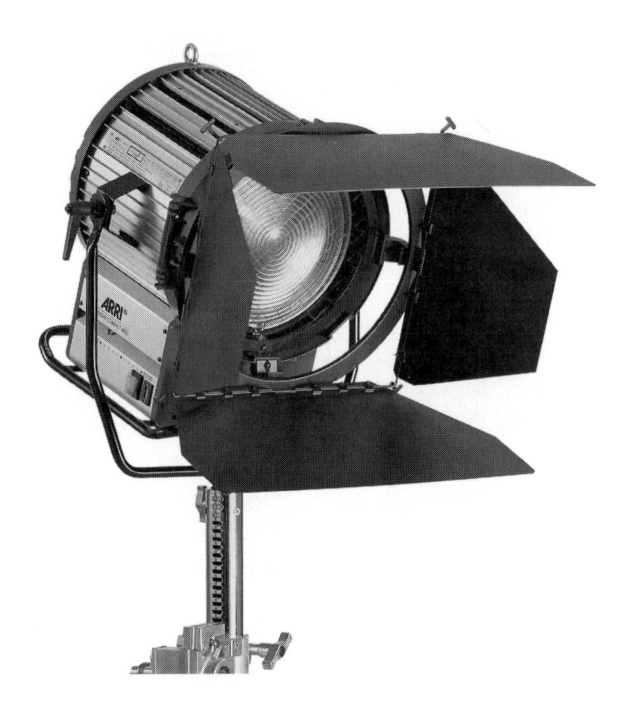

Figure 2-7

An Arri Compact 4000 SB R HMI (Image Courtesy of Arri, Inc.)

#31 How to Use the PAR Light

The output of a 1200 PAR is roughly equal to a 5K tungsten light (not in amps, but actual foot-candles). That's right; a 1200-watt light gives you the same amount of illumination as its larger tungsten cousin (and the HMI doesn't need the power requirement a 5K demands).

Because of their power, PARs are best used when you need a lot of light. I often use them streaming through a window to simulate daylight. Even through window glass, this intense light will look and act like sunlight on your shoot. By gelling the PAR, you can create sunrise or sunset effects by warming or cooling the lamp's output.

Hollywood uses much higher wattage HMI PARs when illuminating a massive set with numerous windows. But on the average room-sized set, two 1200 PARs are sufficient to get the look you're after. Figure 2-8 shows an HMI PAR pounding through a window from outdoors.

These same units can be used indoors as well by placing them in a diffusing Chimera. It is rare for a 1200 PAR to be used indoors without some type of diffusion. As I mentioned earlier, it's like having a 5K on the set (and you would definitely diffuse that). A Chimera not only spreads out the light, but also broadens as well as softens its output. I've used this setup in store locations where the window was providing most of the daylight, and I just needed a daylight-balanced, soft interior fill.

Figure 2-8

An HMI as a key source through a window

In theatrical applications, sometimes you want to simulate a bright white spotlight shining on a subject. If I had used a typical theatrical spotlight, its color balance would be close to tungsten and would not stand out on video; it would just look like another light. But by balancing your camera to tungsten and using an HMI PAR, you now have a spotlight that looks white on camera as well as in real life. Figure 2-9 shows an HMI PAR pretending to be a theatrical light.

Figure 2-9

An HMI acting as a theatrical spotlight

Positioning the PAR 12 rows back in the auditorium, the light should be raised and sandbagged. At a height of 11 feet, this would approximate a balcony spotlight from the camera's vantage point.

#32 How to Use the 1200-Watt Fresnel

The Fresnel's light is modeled by the glass in front. With various textures of glass interchangeable, this HMI creates a distinct look on the talent's face. Sometimes a Fresnel is the only light that will suit a given situation. I've often used Fresnel HMIs when shooting close-ups of talent outdoors where the sun is acting as the key, but I needed a little fill to liven up the shadows. Westerns where the talent is wearing large, brimmed hats are a nightmare with orange tungsten light.

By adding diffusion to the Fresnel, I achieved added texture where it wasn't before. Keeping your color balance in the world of daylight makes your job easier. Most locations are predominately daylight, and having supplementary lighting of a like color temperature saves you a lot of aggravation and unwanted pools of multicolored lights. Figure 2-10 shows a Fresnel on location.

Figure 2-10

A Fresnel acting as a key

#33 How to Use Multiple Cameras with Minimal Headaches

Your problems are naturally multiplied when using more than one camera. Instead of blindly determining when to pan, tilt, dolly, jib, zoom, and focus one camera, sometimes you may be responsible for others (not your children, the cameras). In a recent job as *technical director* (TD) on a fitness video, I had to have a rough idea of what five cameras were doing at all times.

Todd Taylor of Taylor'd Productions was asked to produce an "Absolute Body Power" video for the clever, vivacious, and approachable Alan Harris. He wanted to create a four-tape and DVD set that would keep everyone from Grandma down to Junior in prime physical condition.

Set in a studio at Sheffield Productions in Phoenix, Maryland, Todd arranged for a crew of 25 and a cast of 16 to shoot 2 instructional and 2 workout videos. With the tungsten lighting set on a grid high above the sweating mass below, five Sony cameras would capture the event.

Camera one would offer the wider master shot and follow muscular Alan as he flew to help each student in his trademarked workout. Cameras two and three, both handheld, would capture the left and right angles of the set and jubilee, as well as in-your-face close-ups of Alan as he humorously pleaded for the viewer to work harder and whine less. Camera four was mounted to a 12-foot Jimmy Jib that would swing over the heads of the perspiring hordes as they ran, stepped, lifted, jumped, and wilted. Camera five was bolted to the ceiling of the studio to get a high-angled view of the entire arena. This would be our fail-safe shot if everything else broke.

With an initial duplication run of 50,000 copies, Todd had to hire a TD that knew exactly what he, she, or it was doing; he hired me anyway. Enclosed in a remote truck and looming over a switcher, I had a pristine view of all five cameras on three-inch Sony monitors. In case I cramped when switching, Todd had the foresight to ISO (isolate) each camera to a deck, each with jam-synched timecode. From an editing standpoint, if I misswitched any feed (not that I ever did or would admit to), my error could be corrected in post because of the individual ISO feeds.

Once I was suited up and on my throne, Alan walked right up to camera one and spoke directly to the viewer and the madness began. As I perused each monitor, I noted that each of Alan's students in the front were magnificent specimens of humankind. The women even had six-pack abs (I painted mine on in college). When I regained my composure after drooling over the first row, I noticed the examples in the second and third rows weren't quite as model-like as the four super-models up front. The others were actually real people who could benefit from this high-intensity workout. The men had stomachs and the women had gray hair (only the older ones). The sweat would be real and the grunts lifelike. I was now in for the thrill of a lifetime.

Being that I was involved in an exercise video, all my switching would be hard cuts. The fancy dissolves and diagonal wipes would have to wait for a rainy day. Using licensed music, Alan requested that his "exercites" step it up. In video terms, that means speed up the pitch of the CD. Cranked to Warp Three, fuzz began to fly in the studio; my switching became a four-fingered affair, and the remote truck was rocking (although no one was knocking).

As Alan's physique expanded three inches, he lost four pounds of perspiration through his forehead. With all this fluid cascading on his wireless microphone headset, the unit failed. Like watching a Buster Keaton routine in the late 1920s, Alan jumped around, but we heard very little. With a replacement unit on hand, we were able to add a new mike a few moments later. The average life span of Alan's mike during an Absolute Workout was 23.653 minutes. He wouldn't be invited to share the stage with Madonna in the near future.

Splitting my eyes five different ways, I had to blindly guess where Alan would be next. Would he talk two inches away from camera two's lens like he had in the past? Would I catch the head of camera four's swinging jib in camera one as it climbed for the perfect shot? Would someone step on camera three's CCU cable as she walked into an extreme close-up of abdomens? These questions all were answered the way I thought they would.

I tried to warn each camera operator before I cut to them live. "Ready camera two . . . " Alan would dash over to another workoutee and offer advice. "Take camera three." There's nothing like telling someone to be ready to be switched only to be wrong. Indecision builds strong bones and teeth, just like a workout.

During breaks, I was allowed out of my climate-controlled environment into the hell-like warmth of the well-lit, sauna-baked, sun-dried studio. Opening the front door promising to get the penguins in the remote truck a drink, I was greeted by a swarm of tsetse flies. The air was thick with an intoxicating mixture of body odor and perfume as it grabbed my tonsils with a vice-like grip. I was now seeing firsthand how the other half (the physically fit) lived. Being a lethargic slug has its benefits.

Comfortably overdressed in my long pants and flannel shirt, these poor souls in the easy-bake oven had their clothing spray-painted to their taut bodies. Told not to wipe the sweat off their exuding pours (for continuity) between takes, I managed to dodge the puddles on the slippery floor. Only once did I lose my balance and skate across the room on one toe until I collided with a Sheena Easton look-alike. I apologized for smearing her make-up clothing and politely genuflected away at half-mast. I would spend the rest of my incarceration in my vinyl-covered seat in the truck.

I wondered why all the out-of-shape crew members like me were sitting in an air-conditioned truck while these gorgeous models were working out in the 120-degree studio. They were definitely the models; we were more like the bad examples. I now sweat better from five different angles.

The best advice I can give is to be prepared for anything and keep a finger on each button on the switcher. You'll never know when you will be called into action.

#34 How Video Assist Can Work for You

Even with big-budget Hollywood features, video still plays an essential part. I had the opportunity to flash my video skills to the 35mm world. Relax, that's the only flashing I do.

It seems that the real star of the film, a 1909 steam locomotive, was returning home to shoot some key scenes with people, animals, and animated stars. The following saga is what happened in the video department.

The video assist on the camera's viewfinder is tethered via a BNC cable to a monitor/VCR combo. You plug video out of the video assist on the camera to video in on your monitor. Of course, this really sounds simple; a well-trained chimp could run this setup. Unfortunately, I wasn't a chimp.

I had taken the liberty of gaffer-taping the video monitor screen to mask off all the viewfinder area that wouldn't be seen in the film's 1:1.85 aspect ratio. This also wasn't that difficult to do and it allowed the director of photography and the director to view only the material the rest of the world would see.

The monitors, spare batteries, coiled cable, and AC cords were piled onto my rolling AV cart. Happy memories of wheeling 16mm projectors into classrooms danced in my head; this would be just like the old days, only on video.

Our first shot was going to be a half-mile from the train station. The steam engine number 475 was going to chug out of the station and pass the camera. The 65-member crew packed up and traveled to this remote destination. It was as if a small army of people had gathered together to begin a war.

Once at our spot, I connected by BNC cable to the video assist on the 35mm camera. Since we were far from the station, I had to power both monitors with two batteries each. This, in addition to the recording of the images, would allow me to have approximately 20 minutes of power. Every split second the director turned her back to cough, I would power down my setup to conserve power. This got to be very annoying. As soon as the director looked at the shot and approved it, three grips, two gaffers, and a second assistant camera would walk over to see what the shot looked like. Not wanting to make any enemies, I would switch on the monitors for each of them so they could review the shot. The little monitor was too small to suit most; they all wanted to see the video on the 13-inch monitor. Figure 2-11 shows me standing by my child.

Since there wasn't a cloud in the sky and the DP was shooting at F22, the monitor screens were invisible in the bright sunlight. With a dull knife and a two-by-four-foot piece of posterboard (and six miles of gaffer tape) I created a hood for both monitors. If you rested your eyebrows on the edge of the cardboard and cupped your hands against your face, and if four other people stood next to you, a sharp eye could actually make out an image on the screen.

After 312 people from all walks of life had checked and approved the shot, the assistant director called for the steam engine to begin. Soon afterward, the DAT machine began recording, the slate was inserted, the camera came up to speed, and I hit record on the deck. I had been working my finger out all week so I could press the record button correctly.

Figure 2-11

A boy and his video assist

As soon as the 22-ton steam locomotive reached the center of the frame, the 8mm monitor combo died; the batteries were dead. This immediately cut the image from the bigger screen. With the agility and dexterity of a gazelle, I popped out the two expired batteries. Like a toaster from the 1950s, the springs that held the batteries inside the monitor were rather tight. Once released, the batteries sailed three feet from the deck onto the rocks.

Mopping the cold sweat from my forehead, I inserted the new batteries every which way but the correct one. Finally installed, I prayed that the monitor would power up sufficiently in time for me to actually record the shot (something that I was being paid to do).

As I slammed the red button to the right, the train belched past the camera, spewing a fine mixture of water vapor, unburned coal particles, and a microscopic ash that filled my lungs with a burning itch. When the director yelled cut and looked at me, all the blood left my face. I knew I had only recorded five seconds of the shot before my batteries died. I now knew I was going to experience the same fate as the batteries. If she wanted a playback of the footage, I would have to take the film from the camera and run to the nearest all-day processing lab and explain to the cashier that I didn't want two sets of prints with my negatives. Luckily, the director wanted another take and never asked to review the first shot. I was in video assist heaven.

Shot number two was back at the platform. Within moments, the caravan of film gypsies wheeled and pushed everything back to the station. I never realized how difficult it is to push something on a cart over train tracks and ties. Every time I transversed a tie, a battery would jump off my cart and try to make its break for freedom.

I spent the better part of 5 minutes wrapping my 100-foot-long BNC cable to a fine, tightly coiled loop, only to have it unravel and wrap itself around the nearest grip's feet. I found out that a 100-foot piece of BNC cable can contain no less than 79 knots, tangles, and kinks.

When I arrived at the platform, everything that was on my cart had jumped ship at least twice. Everything had been thrown back on when it fell or I would have been trampled by the herd of nomadic technicians. Luckily, grips are very agile. They are able to push a 400-pound cart loaded with metal stands and still scoop up one of my escapee batteries. They would also do this without stopping, bending, or missing a beat in their cadence. Grips are a video assist's best friend. If they had tails, they would be invincible. Determined to impress someone, I gaffer-taped everything together into neat bundles on the cart; nothing was going to fall the next time. Of course, I didn't count on undoing the gaffer tape at the end of the day; you never have a machete when you really need one.

Three hundred extras had been hired to walk down the station's steps, past the camera, and into the train. I recorded six takes of this action. Each time the director asked for a playback, and I was ready. However, whenever you have a TV set up outside and more than five people are involved, they all want to see it. As the director watched the playback through my makeshift cardboard hood, every extra would slowly inch behind me to see if they could catch a glimpse of themselves on the screen.

I could have been watching *I Love Lucy* reruns on my TV, but somehow they knew better. One time I turned around to find 300 faces starting at the screen trying to see something. Slowly, I was being pushed from my post. I clung to the cart knowing they would have to scrape my dead carcass from the sides. I would never abandon ship! Anytime a person on the set would walk by, they would notice the monitors and ask to see the last shot. Not really knowing who the producer's nephew was, I obliged everyone. I finally caught on when a family traveling on the train to Philadelphia asked to watch the scene. If they had asked for a playback, I would have known they were with the crew. One slip of their tongues and I knew they were phonies.

I would soon learn the real importance of video assist. Every role in a feature film is important, some slightly more than others. As soon as the director yelled cut, all of the big wigs raced over to talk to me. At no time on the shoot was I ever asked to "roll tape."

Everyone else on the set is asked to roll sound, slate, camera, action, and so on. I have to instinctively know when to start and stop recording. The director wants to check the action, the actors want to see their performances, the DP wants to see if the lighting and exposure are correct, the camera operator wants to check that he framed everything the right way, and the first assistant camera wants to make sure he had focused everything correctly. Who else on the set receives that much attention . . . clothed?

In fact, the DP never looked in the camera. He constantly watched video assist. The immediacy of video is necessary. Although it hasn't quite reached the quality of 35mm film, through video assist anyone on the crew can instantly review the shot in its entirety. If, after reviewing the footage on video assist, the shot doesn't measure up, it isn't printed. I actually did save them some money in processing costs. Wait until they get my bill for batteries!

#35 How to Keep Sand from Getting in Your Equipment

You need to remember a few things when shooting at the beach. In case you haven't noticed, there's lots of sand around. The only place you don't want sand is in your camera. When on location, always play back some footage to make sure everything is functioning properly. Back at your office is no place to see if you have any footage recorded.

I've been on beach shoots, and when I've returned, my deck refused to function properly. When a tape was inserted, it wouldn't track correctly when recording or playing. Thinking my tape path had to be realigned, I sent the camera to the shop.

I later learned that *three grains of sand*, tiny little pieces of crushed rock and shell, had found their way inside my camera and were preventing the tape from tracking properly. Digital cameras have such fine tolerances that these three grains were enough to send everything into a tizzy. This was a relatively expensive repair because the camera had to be disassembled and the sand removed.

I never poured or blew sand into the camera. The sand could have entered the camera in only two ways and both of them involved me letting them in. Learn from my expensive mistake and never place your camera on the sand even for a split second. When you pick up the camera again, sand will still be clinging to the bottom. You see this and brush or blow it off, sending the sand who knows where. This is a bad move. If the camera must rest on the ground, put a blanket, towel, or tripod underneath so the sand will not make contact with the camera.

The other possibility was when I changed tapes. Rather than walking 200 feet to the vehicle, I popped out the first tape and inserted the second that could have been contaminated from my hands (which were used to wipe sand off the bottom of the unit). Make sure your hands are clean and reload tapes someplace other than the beach. This takes a few minutes longer, but you will save yourself a lot of money in the long run.

Just remember, keep the pretty sand as far from your camera as possible. It's small and fine enough to get where you don't want it (remember the last time you got some in your suit?) and basically it's sandpaper without the paper back. It's better in Bermuda; at least their sand is pink.

Chapter 3

Production Software Solutions

#36 Production Software Solutions

In addition to the production hardware is the software solution, or letting your computer do the work for you. This chapter will discuss how to choose the best *nonlinear editing* (NLE) system, how to save time and money by using logging software, and how to use effects that happen as soon as you choose them.

#37 How to Find the Best NLE System

NLE systems have all but changed the course of editing as we know it. Gone are the days when you electronically lifted one shot from tape A and placed it on tape B with a second shot, only to find that you must add a third shot between shots one and two. Previously, that meant erasing shot 2, adding shot 1A, and then reinserting the old shot 2, which is now 3.

It may sound confusing, but it's really not. With an NLE system, shot 1A may be placed between shots 1 and 2, with the new image just moving shot 2 farther back. If you determine later that shot 1A doesn't work, it may be removed and the gap where it once resided is closed. This immediacy and changeability is what makes an NLE system indispensable when doing any kind of video editing.

But how do you decide which system is best for you? The answer to that question is to ask yourself more questions. What kind of editing will I be doing and how often? An event videographer may do less editing and need less features than a documentary or feature editor. How much information will I be editing (TV commercials, documentaries, or longer feature films)? This lets you decide how much storage space you'll need. How rugged does this system need to be and will I be taking it to locations or will it reside on a desktop? If someone has to approve everything you do, a portable system might be more accessible. You can ask many more questions, but depending on your specific needs, you can find systems out there that will fit the bill.

Presently, more than 30 different NLE systems are available and it may be difficult to determine which one best suits your needs. I'll break the NLE world down into three classes: entry level, midrange, and high end. Each system within these categories is excellent; some just offer more frills and features at a higher price tag.

#38 Entry-Level NLE Systems

The best feature about an entry-level system is its cost. For under $500 (and a computer), you can begin editing digital video. However, you won't have certain capabilities; these may not be things you are concerned with for your specific edit projects, but they should be discussed. The entry-level system manufacturers include Applied Magic ScreenPlay Plus (www.applied-magic.com), 1 Beyond 990 System (www.1beyond.com), Canopus DV Raptor-RT (www.canopuscorp.com), Laird DVC1-XPDV (www.lairdtelemedia.com), and Strata DV-Pro (www.strata.com). To make things even easier, I'll further break down the entry-level NLE systems into two subcategories: real-time versus rendered effects and input options.

#39 Real-Time Effects

Later in this chapter, I'll explain how to best utilize real-time effects, but a simple definition is "any video effect that happens immediately without the user having to wait for it to render." Most entry-level systems do not offer this luxury and you must wait for any effect to render. A one-second dissolve might only take three seconds to render, but if you have a video with 361 of these dissolves at three seconds a piece, that adds up. In order to keep the cost of the software down, some NLE systems do not have effects that happen in real time. If you have short deadlines, these NLEs will take more time.

#40 Input Options

What kind of video do you want to input to your editing system? All entry-level NLE systems enable you to send digital video only directly into your computer, usually via an *Institute of Electrical and Electronics Engineers* (IEEE) 1394 input (commonly called FireWire). This way DVCAM, Mini-DV, DVC-PRO, and Digital 8 images can be sent directly into your system with one cable that handles timecode, video, and stereo audio.

If you have an analog camera, you must purchase an external box or card to input the video signal (via a BNC, RCA, or S-VHS connection) as well as left and right audio channels. Entry-level systems sometimes offer this but at a higher price (usually at least $300), which puts us into the next category anyway.

#41 Midrange NLE Systems

This group ($500–$1000) is the largest sector and includes Abode Premiere (www.adobe.com), Avid DV-Xpress (www.avid.com), Apple Final Cut Pro (www.apple.com), Pinnacle Cinewave (www.pinnaclesys.com), and Matrox DigiSuite Max (www.matrox.com). These systems include real-time effects and more features than the entry-level grouping.

I've used every NLE system listed previously and found each to have its own strengths and weaknesses. Learn all you can about each one from the web sites, product literature, reviews, salespeople, and editors who have used them.

#42 High-End NLE Systems

This grouping (over $1000) contains the more "professional" offerings and includes Discreet Fire 5 and Smoke 5 (www.arrow.com), Media 100 844/X (www.media100.com), Avid 8000 (www.avid.com), and Thomson Grass Valley Broadcast Solutions News Edit (800-824-5127). For people who edit for a living, these somewhat expensive systems leave no stone unturned, or basically there is nothing these units cannot do. A colleague purchased a $60,000 NLE system, and within two years a system costing $1000 took up less real estate and mirrored the behemoth's capabilities.

In short, no one can tell you which system to purchase or lease. Try each one out and see which is easiest to use (the learning curve), suits your needs (handles the type of editing you'll be doing), and fits into your budget (the versions will change often and the price will come down). With a precise NLE installed in something as small as a laptop, the next Academy-Award-winning picture could be edited on the computer you're using now.

#43 How to Use Logging Software

As mentioned in Chapter 1, logging software now takes the mundane tedium out of handwriting each shot number, description, timecode in and out point, and any other information. Computer programs now do it faster and more accurately. You have no way around it; every shot you shoot must be logged.

Of course, this is one of the most mind-numbing things about video production. When an average shoot necessitates logging hours of footage and sorting it into a useable form that can be accessed by all interested parties, I am always yelling at the logging person for falling asleep and turning in massive gaps with no information.

Logging is an art and must be precisely done. The Executive Producer® (TEPX™) Silver Edition by Imagine Products Inc. has made this tedious job much easier. This isn't going to be a review for the product, but it's one of the two logging systems currently out there. Unlike most NLE systems, TEPX works with any DV camcorder without asking you the make and model or putting you through a lengthy setup.

Having shot nine hours of mini-DV footage, I needed to review and log all the images and allow my client to offer his two cents in selecting the best takes (and approve them). With four viewing modes, Entry View, List View, All View, and Edit View, you can choose the logging template you desire or custom design your own.

In the Entry View mode, a tiny video screen captures a JPEG still image and marks the in and out points and durations automatically, allowing the user to enter comments, project information, dates, and so on. Repetitive information may be automatically logged in to save your fingers, and TEPX even has *Auto Capture Technology* (ACT) with user-selectable sensitivity, which will log your footage by entering the in and out points on the fly. I'm told the software senses the starting and ending points of each shot.

Using the small viewing window, you can watch and hear all your footage in color and at 30 frames per second without the video skipping or stuttering, just like you would if you were logging it in the field and watching through the camera's viewfinder or monitor. Maybe I'm easily amused, but having a color still image of my footage is just like editing in an NLE and I avoid handwriting log sheets. The TEPX enables you to store shows in a library that can be called upon later (38 edits down the road) and that one log will have all the pertinent information without ditto marks that are meaningless alone.

It's also extremely easy to let the TEPX grab your JPEG frames and mark points, knowing you are building a batch digitizing list in the process. This may be accomplished automatically, manually using the mark in and out points, or via keyboard shortcuts.

Using the Auto Fill function, you can keep everything with the same project name without having to retype the same phrases over and over. I've never had the luxury of a spell check on a logging program, but my logs are spelled correctly, and a production assistant of mine had no trouble learning the system in five minutes. Like an NLE system, shots can be rearranged, sorted, changed, added, or deleted at the user's whim.

Exporting your completed log to any number of NLE systems is also a breeze from the oldest ASCI to the newest Final Cut Pro. It's nice to know this logging works on some antique editing programs.

I print out my completed logs and have a hard copy by my side during the shot selection process. With numerous audio takes, just logging in the beginning and ending sentences of text can save you hours and lots of money in transcribing fees.

In film school I was told that the key to being a great editor is organization. No fantastic editors are disorganized slobs (I'm not fantastic). With logging software, you can't help but be organized. It does all the mundane tasks for you and makes you look organized in the process.

#44 How to Use Real-Time Effects

In this fast-paced world, editors are requesting real-time effects in their NLE programs. The old adage used to be, "Edit your program, select 'Render,' and then grab a cup of coffee because the computer has to do its thing for a while." This faster real time, however, comes with a price; real-time effects not only add to the cost of the system, but also suck up your computer's memory in order to display those effects on demand.

If you have the memory, a real-time effect will greatly speed up your editing time, but bog down how fast things happen on your screen. With most computer editing systems having 1 gig or more of memory, this is rarely a problem; it's those with only 256K or 512K that experience problems. When working on a project with numerous effects, you can choose to preview these effects instantly (real time), save the memory space and have only complex effects happen in real time, or edit on the fly and have all effects happen in real time later when you output your *edit decision list* (EDL).

With Avid, for example, more than 100 infinitely customizable real-time effects, including compositing, color correcting, and titling, are integrated with the software. More complicated effects like real-time color correction will slow down your editing process because of the RAM needed for this effect. During the rough-cut phase of your project, it's sometimes best to save these fine-tuning or sweetening effects until the final run because of the RAM needed. Bogging down your computer for final tweaks is best left for the end. Instead, spend your time on assembling the program with wipes and dissolves as the only real-time effects activated.

The best question to ask yourself with a real-time effect is, "Do I really need it instantly, or would I be better suited to view the completed effect at a later time?" If you can answer yes to this question, save the memory for faster processing time in editing rather than having all the flesh tones match from shot to shot.

Chapter 4

Working with Extras

Solutions in this Chapter:

#45 How to Work with Extras

Times will occur when you'll need extras to mill about in the background. These people don't have any lines to remember; they just have to be in the right place when the camera starts to roll.

But, like cattle, are extras really a dime a dozen? The fact is that for anyone who wants to be in a video, this would be a great way to begin. This will give them their 15 minutes of fame and it pays them a few bucks in the process.

A feature production crew rolled into town and they desired 150 extras to fill a railroad station. Because this scene involved Thomas the Tank Engine, they believed every child on earth would clamor to be in the locomotive's company. The only problem was that the shoot occurred on a Friday in September when most kids above the age of five are in school. The casting director (from Toronto) asked me, because I was a local boy, if I could contact some of my friends to be extras. I told her I knew lots of kids and asked how many she would need.

When she told me she wanted 150, I realized if I knew that many kids, the Internet police would be watching my every move. She also told me she would feed the children and their parent(s) as well as give each child $50 for the day. Giving a kid $5 an hour to stand around and watch their favorite nonpurple creature would be easy, but having a parent miss work and share in that same dubious glory was a little harder to swallow. Nevertheless, it was my task and I was going to make it happen (until I hit the first hurtle). In addition, she also needed a few parents to hang around and be extras in the early evening.

#46 How to Use Child Extras

I now had the dubious honor of finding a hoard of kids, paying them more money than they've ever seen, and separating them from their parents while Mommy pretends she's a movie star; this was going to work. It's extremely easy to find people to be extras. Everyone wants to do it—until you explain all the work (or standing around) that is involved. However, the mystique about any kind of film or video is hard to see through. Once again, don't pretend that your project is anything it's not. Be truthful and explain their role and how it will help the project.

It seems that the kids' will was strong, but the flesh (their parents) was weak. But I figured if I planted the seed and the kids whined enough to their parents, I'd have them begging me to take them off their hands for the day. I did not want 150 children tagging along and following me over the course of a day, but I still had a job to do.

I asked every parent I knew who didn't have a child in daycare (three of them). Most, however, were slightly more interested in what their role would be. "Of course I want what's best for little Pricilla, but how big is my part?" they would casually ask. I had a difficult time getting kids, but an easy time getting their parents. I finally contacted a school principal and explained what we needed and asked if that could be a good field trip or class project. After several phone calls to numerous schools, I found the right number of miniscule extras who would be bussed to the location. Figure 4-1 shows how many people "a few extras" actually is.

Figure 4-1

A few extras on the Thomas shoot

#47 How to Use Adult Extras

I explained that each parent would pretend to be an executive running around the train station. Each would be given a dummy cell phone and they would walk in a predetermined direction, pretending to be gabbing on the electronic brain scrambler. The prop guy had a box of over 100 cell phones whose contracts and activation times had expired. When the box had emptied, he switched to wooden blocks painted black. At a distance, no one could tell if you were talking into a piece of wood or a real, dead cell phone (is there any difference?).

If it ever becomes your job to get extras for the director, you will be responsible for them. Like saving someone's life, they will be indebted to you the rest of the day. Because they "know" you and you got them this Hollywood-type gig, they will follow your direction before the real director's. This behavior must be nipped in the bud. Tell them you will be busy elsewhere, but to listen to the director and do what they are told. Once they are aware that they will be getting paid, suddenly your contacts are now professionals. They will be given (or told) what to wear, when to do something, and when they can eat.

Sadly, you must burst their bubble early on. Most will be standing around a lot doing nothing but waiting. Some may not even be used on camera. They cannot wander over and talk to the stars about how much they love them. Start by telling them the things they shouldn't do, and then explain how much fun they will have with the experience. They will forget all the standing around when they tell their friends that they were in a movie. It doesn't matter if it is a big budget feature, a corporate video, or a TV spot. They appeared in the video and were in front of the camera.

"Yeah, Ralph, I spent the day on the set with Mr. Hackman. We chatted at lunch, but I was too busy to really pay attention to much else. Once the film gets released, I'll have to consider if I want to do this full time." If you were to translate what really happened, it might go like this: "I spent the day with Bobby Hackman, the second assistant director who told me where to stand. We extras sat together on the grass and had a boxed lunch. The video is going to be shown to the corporate CFO before the employees see it during their break. They told me it was a pleasure working with me and we should do it again sometime."

Don't break the myth; let them believe what they want. It will be a day they will never forget (a boring one). So pay them for their time (the amount isn't as important as the opportunity they received), give them a copy of the finished work (no matter where it does or does not air), and thank them for being such a big help.

#48 How to Feed and Care for Extras

In order to win friends with the extras waiting in line to see Thomas between takes, I took leftover bags of potato chips and sodas (from our lunch) and passed them out. If I had given them gold, they could not have been happier. As I gave the freebies out, some asked, "How much?" Others asked, "Why?" with a sly look in their eye, and others ripped the bag from my hands without saying anything.

This may be just another day in your life, but everyone who is a nonprofessional on the set is in seventh heaven. Do whatever you can to make the day more enjoyable for them. Most of your day will be full of duties, but any crumb (or bag of chips) that you can pass along to others will make it a memory for them that will last forever.

Food is one of the cheapest and nicest things you can provide for any extra. As I mentioned earlier, everything about this day is different from their normal routine. Some will remember one specific thing more than another about the day, but if you do not feed them, *that* is what they will remember. Keep everything positive about the day. A little food goes a long way.

#49 How to Get Actors While on Location

I had a shoot on a tropical island in the Caribbean where the client didn't want to pay thousands of dollars a day for beautiful people to frolic on the beach, dine on gourmet food, or run in slow motion. It was now my job as the producer to get "free" people to appear in the travel DVD.

Guests at the resort were thrilled for the opportunity to share a lobster in front of my camera. I told them we needed to shoot a couple (I did say film, not shoot) having a lobster dinner and asked if they would be willing. My first choice actually said no, but the second couple agreed. I shot the scene over the course of five minutes as they laughed, talked without saying any words (they moved their lips), and pushed the lobster around on their plates. After I got the shots I needed, I told them to dig in and enjoy. Their only payment was the free dinner, I needed to videotape it anyway, and I disposed of the lobster in the process. Figure 4-2 shows that guests can be convincing actors.

Even in a tropical paradise, you are taking people away from their time together, their honeymoon, the Witness Relocation Program, or their vacation. Help make it even more memorable for them by giving them something and a copy of the final product for their time.

In whatever situation, and whatever gift you give them, if you use their likeness, they must sign a release. Even if they kiss your feet afterwards for

Figure 4-2

A couple sharing a free meal

shooting them kissing under the stars, they can turn around as sue you because she later noticed she has a piece of barley stuck in her teeth and the world saw it. Once the release is signed, your butt is covered and you can now continue your search for the perfect person. If you ever find that elusive person, let me know.

Solutions in this Chapter:

Chapter 5

Using Consumer Digital Cameras

#50 How to Use a Consumer Digital Camera

If you want to get into digital film making, the cost of the camera shouldn't be what's keeping you from starting. In this chapter, you'll learn how to make an inexpensive one-chip camera work for you rather than against you; how to best utilize an image stabilizer (whether it's optical or electronic); making an inexpensive lens work and look like one costing thousands more; and, with the myriad of digital tape formats available, how to choose the one that's best for you.

#51 How to Make a One-Chip Camera Work for You

Not everyone can afford the latest three-chip professional camcorder. Some only have the money to purchase an inexpensive single-chip model. As I mentioned earlier, the price of the camera should not stop you from making videos. But how can you get professional-looking results from a camera costing around $500?

The one word that comes to mind is *lighting*. With professional lighting (indoors or out), an inexpensive camera will look much more like its three-chip cousin. When video became available to the consumer, a one-chip camera was light-years ahead of the obsolete tube camera. With the addition of the CCD, the camera could be pointed at the sun or a bright object without worrying about the excessively bright image being burned into the tube. Other things like lag, smearing, and comet tailing became things of the past when CCDs became the vogue.

(Continued)

#61
Choosing Mini-DV

#62
Choosing DVCAM

#63
Choosing DVC-PRO

The tube camera could even fail if it were pointed towards the ground for extended periods of time. Being almost indestructible, the new chip cameras can do almost anything. But with the public wanting more, a camera with three chips could do so much more than one. The main differences visually (besides the cost) are that the separation of color and sharpness is better in a camera with three times the number of pixels. A one-chip camera with a 410,000-pixel CCD chip won't look as sharp as one with three 410,000-pixel CCD chips.

You cannot match the clarity of a three-chip camera with one that has one-third of the pixels, but you can enlist it in the army and make it "be all that it can be." Getting as much light as possible into the single CCD will help make the image look its best. Under low lighting conditions, the grain will be more accentuated in the lesser cameras. By taking your camera outside or using additional artificial light to boost the gain, your image will look its best.

Every camera has its so-called sweet spot where it performs optimally, and that is usually at the highest f-stop numbers (F8 and above). By correctly exposing your image and using that f-stop range, your image will look better than one at F1.4. The colors will still separate slightly and some grain will be present, but you will be getting the most out of your camera.

An image shot in the telephoto mode will compress the image and have fewer problems than the same image shot in wide angle. Too much information is captured and details will be lost when shooting in wide angle (just as a wide-open f-stop degrades the image).

Avoid using the digital zooms on consumer cameras that magnify 360 times and up. If you look at an image with digital magnification, beyond a certain point it looks too pixilated and grainy.

Shooting on digital tape allows you to record the image in the best of all possible worlds. The same one-chip camera recording to an analog format won't look as sharp or clear as the same number of pixels do to a digital medium because the tape is formulated better and won't be marred by dropouts (loss of oxide).

By being professional behind the camera, a nonpro camera will perform better than if a novice is using it. Little things like a steady shot (use a tripod), avoiding overzooming (too much zooming looks amateurish), a high f-stop (the larger the quantity of light, the sharper the picture), and shooting outdoors whenever possible makes it more difficult to tell the one chip apart from the three chip.

Practice with your one chip and learn its limitations. What shots look best and which don't look quite as good? Every camera is different and once you learn yours, you will get the most from it.

#52 How to Use Image Stabilizers

Those of you that need assistance with keeping your camera shot steady have been aided with the addition of a camera's image stabilizer. I've suffered trying to handhold a camera for over an hour, and the stabilizer has taken most of the movement from my shot. This is a great feature and most of us have taken it for granted.

Under no circumstances should you believe that a stabilizer will solve all your shaking problems. A tripod is the only cure for a truly steady shot. But before the advent of the image stabilizer, fine movements like breathing would be telegraphed to the camera and be visible to the viewer. Look at what happens when you are holding the camera and waiting for Junior to take his first step; it could be an hour or a day. When the camera isn't turned on, he's walking all over the place, but as soon as he sees the camera, it's now the waiting camera and you are left holding it.

Stabilization is meant to assist but not totally compensate for camera movement. If you follow someone as they are walking, the stabilized image will still show a slight bounce as you make each step (which is natural), but much of the unevenness has been eliminated. If you try the same tracking shot while the subject and you are both running, the stabilizer will have no effect; the camera bounces all over the place. Slight side-to-side and up-and-down oscillations are controlled, but that's about it.

You should keep a few other things in mind when using an image stabilizer. It consumes a lot of power, but this isn't a problem if you are shooting with the power cord plugged into a wall. If you are trying to get a shot with a nearly dead battery, however, it could be something to be aware of. If you are using a tripod, turn the image stabilization off because it degrades your recorded image slightly. Two types of image stabilizers are on the market: optical and digital (electronic).

#53 Optical Image Stabilizer

The optical stabilizer is the most expensive and sophisticated type of image stabilization. When using this type of device, the stabilizer produces a more even, smoother look to the shot. A small gyroscope is incorporated into the lens system to steady the image. Like a Steadicam or Glidecam unit for stabilizing cameras, the same principle of the gyroscope produces a much smoother image. These are usually found on three-chip cameras and most of these are in the professional line because of the expense involved in the manufacturing process. If you happen to find this on a lower-priced camera, it should definitely be a feature to help you consider the purchase.

#54 Digital Image Stabilizer

The digital or electronic stabilizer uses an electronic circuit within the camera to accomplish the image stabilization. The combination of the circuit and software functions hand in hand to help smooth the shot for the user—at a price. The cost to produce an image stabilization system is far less in the digital realm than the optical. This is why almost every digital camera made incorporates this version rather than the more expensive (and better) optical.

The side effects of the digital stabilization system are some strange artifacts that show up in your recorded image. If you are shooting something that doesn't need image stabilization (the camera is on a tripod or some other fixed surface), turn off the feature. You will see a slight improvement in the picture. The digital image stabilization electronically affects the image and any "enhancement" never really looks natural (just like "simulated" or "electronically enhanced stereo" never sounds as good as the real thing).

However, if you need stabilization because the image is jittering, leave the feature turned on. It's always wise to experiment with it on and off and notice the difference between the two. That is why it's always a good habit to disable the function when using a tripod; you will get a sharper picture and you don't need the option anyway.

#55 How to Make a Cheap Lens Perform Like a Professional Lens

Smaller camcorders don't have the high-quality lens or options of their professional cousins. But with some knowledge, patience, and skill, you can get the most out of your consumer lens.

What really separates a good camera from a great one is the glass in front. The CCD size (1/4", 1/3", 1/2", and so on) and the number of chips (one or three) also play a role, but a lens costing $15,000 will produce better pictures on a cheap camera than an inexpensive lens on the same camera (if you can get it to fit). The precision of the optics and the clarity of the glass are what makes an expensive lens worth its cost. Cheap lenses function just like professional lenses with a few exceptions: the zoom capabilities, digital magnification, and the focal range.

#56 Zoom Capabilities

Several facets make a zoom lens all it's cracked up to be (bad choice of words). The zoom capability is one of the greatest differences. An expensive lens will travel from wide angle to telephoto through its zoom ratio. This is usually referred to in a number ratio, that is, 8 to 1, 10 to 1, 12 to 1, and so on. The range from widest to most telephoto will be greater in an expensive lens, but all digital camcorders have a decent zoom range of at least 10 to 1, meaning that the most telephoto image is 10 times that of the wide-angle image.

The zoom rate (going from one lens setting to another) is controlled by a toggle switch. When you press W, the lens zooms to its widest angle, and pressing T sends you to the opposite end. It is nearly impossible to feather a zoom (zooming in or out very slowly) on a cheaper lens. The motor and gearing ratio isn't good enough to allow you to creep into a close-up.

To make that cheaper lens work for you, forget about using the power zoom. If you are going from a close-up to a wide shot, simply cut from one angle to another. If you try to zoom, the speed or jerkiness of the movement tells the world, "I'm using an inexpensive lens." Rather than calling attention to that fact, don't use the power zoom.

The range of the zoom isn't as long with a cheaper lens. Instead of keeping the camera at a preset distance and using an expensive lens to zoom into that close-up, use your cheaper lens and move the camera closer to the subject. You've just wasted 12 seconds of your time and saved yourself thousands of dollars (a slight exaggeration, but you get the point).

#57 Digital Magnification

Have you noticed that all cameras have a 200x, 300x, or 400x digital zoom? That means once the telephoto has reached its maximum, the electronic digital circuitry magnifies the pixels and brings you in even closer: the higher the x number, the greater the magnification.

If you've ever played around with this feature, even on an expensive camera, anything greater than 10x looks like monkey vomit with all the enlarged grain and pixilation. The image is just too blown up and distorted to be believable. The easiest way to make your cheap camera perform like an expensive unit is to switch the digital zoom off and move the camera closer. The image will look better and you can use the exercise.

#58 Focal Range

The focal range is how well you can focus the image over a specific distance. All digital cameras have auto-focus. Do yourself a favor and switch this off. Even three-chip cameras have this feature and under most conditions the auto-focus isn't going to know what you're trying to focus on. The telltale sign of this is that the lens searches for an image. You have noticed this constant "in focus, out of focus" back and forth movement of the lens accompanied by the whirring noise of the motor. This is usually most evident on a moving shot.

Make your inexpensive lens excel by focusing manually and forget the auto mode. Now no one will ever know that you don't have an expensive camera because you have eliminated all the telltale signs.

#59 How to Choose a Digital Tape Format

Digital is here to stay and I believe that the tape format will also be around for quite some time. But how do you decide which of the digital video formats to use? Luckily, you don't have too many choices, and each has its strengths and weaknesses. I'll discuss each one in detail beginning with the least expensive (Digital 8), traveling through the midrange options (Mini-DV and DVCAM), and stopping with the most expensive in the camcorder format (DVC-PRO). I won't be discussing the high-end broadcast digital offerings such as Betacam SX, Digital Betacam, and High Definition because these cameras begin at $40,000 and escalate from there. These formats are the best, but out of range for most people (the tapes cost a fortune too.)

#60 Choosing Digital 8

A hybrid of the Hi8 analog format, this tape is no different in makeup or price than the old 8mm offering. The only difference is that the tape, normally two hours in Hi8, runs at twice the speed in Digital 8, offering only one hour from the identical tape. These cassettes are the cheapest in the metal particle arena and may be purchased for as little as $4.

This inexpensive format has its pros and cons. Being a nice guy, I'll discuss the pros first. If you need a digital camcorder and don't have a lot of money to spend, this may be your choice. It boasts everything good digital has to offer: FireWire input/output, no dropouts, digital clarity and resolution, 60 minutes of recording time on a single cassette, readily available tape stock in any store at a low price, cameras in the $500 range, and everything else that makes digital shine above analog video. In fact, when Hi8 cameras were falling from popularity, the advent of Digital 8 brought tape stock into stores that never carried it previously. You could also play all your old home videos recorded on 8mm and Hi8 in the same camera that offers a digital output (FireWire) to your editing system.

The two major cons are that Sony is the only manufacturer that offers this camcorder. They have several cameras in their lineup, but if Sony pulls out, this format will fade into oblivion. The other drawback is that the camera is the only means to record and play back your images. All other digital formats offer decks to play back tapes, saving wear and tear on the delicate camcorder drive mechanism. If anything happens to your camera, you have no way of playing back any of your tapes.

This planned obsolescence makes the Digital 8 format one that will disappear the same way videodiscs did. But if you want a cheap digital camera that will play all your old home videos, at least check it out.

#61 Choosing Mini-DV

By far the most popular digital camcorder format is Mini-DV. Every camcorder manufacturer makes a camera that uses this universal format. With prices ranging from around $400 to over $4,000 (in the three-chip models), this format will be around for a long time.

The pros include universality, in that the Mini-DV tape plays and records in hundreds of cameras and the tapes are available everywhere. Other good points include one hour of record time on a cassette costing around $6; two recording speeds (SP and LP), the latter offering 90 minutes on a single cassette; FireWire in and out; and the same tape may be used in broadcast cameras delivering higher resolution and clarity on the same cassette.

The only con is that the quality is infinitesimally less that the next candidate. With millions of these camcorders in existence, I think you know what my personal choice is.

#62 Choosing DVCAM

DVCAM is the grownup (or professional) brother of Mini-DV and a Sony creation. The cassettes and tape are identical, and just like Hi8 and Digital 8, the DVCAM system runs the tape slightly faster, allowing only 40 minutes of record time on a plastic cassette.

The pros of DVCAM are that, because the tape is running faster, the image quality is better. This is because of the fact that DVCAM is offered mainly on three-chip cameras. From the onset, you have a better camera, a more professional lens, and three chips recording to a digital tape; thus, the image will look better. In addition, DVCAM is available in larger cassettes with 124- and 184-minute lengths. A feature film can easily be shot on a few of these longer-format tapes costing between $40 and $60 respectively.

The only con in this format is its delicate drive mechanism. This is a tiny tape that does the same work as its larger cousins in the broadcast field.

#63 Choosing DVC-PRO

DVC-PRO is Panasonic's offering in the world of digital tape. Much like DVCAM, the tape comes in smaller 40-minute cassettes, as well as longer 2- and 3-hour versions.

The makeup of the tape is identical to DVCAM, but Panasonic scans the image onto the tape in a slightly different way. I do prefer the bright, yellow-colored DVC-PRO cassette over the dull gray of DVCAM.

The only drawback again, like Digital 8, is that Panasonic is the only company that supports this format. It's a great tape from a great company, but it would be nice to know you are not the only kid on the block with the camera and tape format.

#63

Chapter 6

Location Shooting

Solutions in this Chapter:

#64 How to Shoot on Location

A location is defined as any place you currently are (that's my definition; *Webster's* says it slightly better). When you are away from your environment (office, home, or set), you are "on location."

You must deal with quite a few things when you are out of your natural element. How do you move a camera from place to place and keep everything consistent? What if you are shooting in the same place week after week but are using different cameras?

Sometimes you must deal with difficult locations like shooting in the snow, at the beach, at sporting events, or at places where children scream and yell, such as an amusement park. Each place has its own problems; let's see how to solve them.

#65 How to Move Cameras from Location to Location and Maintain Consistency

There's nothing like the joy of setting up for a shoot, working feverishly to capture the best moments on videotape, and then sadly dismantling your setup. This is what most people do on every location shoot. But what happens if you have to shoot in the exact same location again tomorrow or next week?

Sometimes you don't have the luxury of leaving your equipment setup for a future shoot. The space you used may be needed for another event, so you must pack up your toys and go home. But your real desire is to keep the cameras in the same spot so the shots will match. This is what makes portable equipment a real timesaver.

If you're using more than one camera, a master control setup should be staged (out of sight) to switch the images. The key in a setup like this is mobility. Everything needs to be exactly where you want it and then be torn down after the show.

We needed to record a 100-member choir who filed onto the stage for a practice run before the performance. Since this was our only chance at rehearsal, the director would switch the images live onto the room's screens. The weight from the singers pushed the cable deep into the fibers of the carpet. Although all cabling was gaffer-taped securely, the body heat from the singers fused the tape into a molten mass of sticky goo. It was now one with the carpet.

Five minutes after the final event was over, the equipment had to be moved to another location because the room was being set up for a new event that didn't require cameras. Like weary gypsy sojourners, we traveled from room to room setting up our cameras, recording the event, and then moving to another room to repeat the process.

Setting up for an event like this takes preplanning. I suggested marking the locations with Xs on the floor, but when we left the room, we would take the Xs with us and put them down in the next room. It reminded me of the time when I marked my favorite fishing hole with taping an X on the front of the boat (I guess I'm not too bright).

Everything is in the timing. Try to have your setup and teardown streamlined so they occur as quickly as possible. The more you have bundled together, the faster it will proceed.

#66 How to Shoot in the Same Place Each Week Using Different Cameras

A local ministry arranged to have 13 church services broadcast. Two episodes would be shot each week back to back at the church and then edited for air. This necessitated a little preparation. Unlike the situation when the cameras had to travel from room to room, in this case the room would stay the same shoot after shoot, but the cameras would have to be removed. Budget constraints necessitated having cameras set up in the church and the master control room being located in one of the nurseries.

After both services were completed on Sunday morning, all the equipment had to be removed before the Sunday evening service and reassembled for the next Sunday morning in the same locations. This constant setup and teardown needed a game plan devised on paper before the first magnetic particle on the tape was rearranged.

The key is to have a quiet, separate area to be called master control. Although the bathrooms have the privacy, running water, and great acoustics, you get too much traffic during certain times. The director, Todd

Taylor was in constant contact with the crew via the headsets, but he needed isolation to concentrate on what was at hand (another reason why bathrooms don't work).

We devised an inexpensive setup that could easily be assembled and removed nine times with no evidence remaining. The cameras, *central control units* (CCU) units, headsets, and the switcher were rack-mounted in a cartable case. The deck was within easy reach, as was the 12-inch preview and program monitors.

An audio board, with 10 microphone inputs and 4 stereo line inputs, was implemented to record the live music and the pastor. This was the only piece of studio equipment that was out in the church; everything else was in the nursery. The sound equipment was mature enough to be in the auditorium, and the sound had to be mixed for the PA system anyway (the operator could manage both at the same time).

Camera one was tripod-mounted on a 10-inch-high plywood riser to capture the pastor (we let him go after we caught him). Camera two was in the front of the church to record cutaways of the congregation. Both cameras were cabled to CCU units that allowed shading control (controlling the camera's iris).

Mount your moveable equipment onto a mobile rack or on wheels. This allows it to be brought in and removed easily. Many manufacturers make cases for transporting or storing the valuable merchandise between shoots. Ours was safe in the nursery because the print was too small for the kids to read and no stickers of purple dinosaurs were on the cases.

Finally, I was allowed to put my gaffer-taped Xs on the floor because the cameras would be in the same spot, in the same room, each week. The X on the plywood riser was removed because helicopters might want to use it, but the other marks stayed.

Tape everything down that doesn't move. If people are crossing over cables, they will trip and fall. When cleaning up, remove all tape from cables and save it in a large ball. It's safer than snow and it helps remove lint on your friends.

No matter what the circumstances, make the best of your mobile situation. Remember, it's not the equipment that makes the show; it's the talent behind it. Pretend you're traveling vagabonds for a circus. Just don't sleep near the elephants.

The key when doing something week after week is to mark what should be marked (find out what works and what doesn't), keep everything neat and tidy (the cables should be wrapped, bundled, and not just coiled in a pile), and rehearse as much as possible before you record. Churches pose their own set of problems because of wood and stained glass, but that's a topic for another Sun-day. (If I spelled it Sunnyday, you would have said I couldn't spell, but the pun might be easier to understand.)

#67 How to Shoot in the Snow

Shooting in snow is one of the greatest challenges for any aspiring videoperson. You have extremes in contrast that are found nowhere else on earth (even at the beach). So how do you comfortably videotape someone when your f-stop is F45 and everyone is wearing dark clothing?

Whenever I have the luck to shoot in snow, the sky is cloudless. Evenly diffused light (an overcast winter day) is best because the contrast between the snow and its surroundings is less. But if the sun is shining, in order to successfully record an image, the bright whiteness of the snow must be knocked down a notch or two. You can wait a week until the snow gets dirty, or you can use a polarizing filter to cut some of the glare. Rotate the filter on your lens until you have the best image, but this will only do so much.

A *neutral density* (ND) filter on your camera is also imperative. Since depth of field is seldom a problem in vast areas of white, use the darkest ND filter you can find. This won't change the whiteness of the snow, but it will lower your f-stop by at least two stops and give your video camera an easier time. On some camcorders, a backlight button is another possible option. This just opens your iris one stop; you can do the same thing yourself.

The shutter speed should also be raised to 1/2000th of a second. Like a still camera, this will increase the amount of light needed to capture an image but with a slight side effect. With a shutter speed this fast, any action in the shot is frozen (pardon the pun) in time. Movement has a tendency to strobe slightly at this speed, but that can work to your advantage.

I did a snowtubing commercial where we used the previously mentioned shutter speed. People did strobe when they went down the mountain, but the snow spray would suspend in time. This became an interesting effect that the client loved. How often can you freeze particles of snow (only in winter) and have them remain floating in time and space for a brief period?

It's best to let the snow be overexposed slightly while the people involved are exposed correctly. If the snow is two stops too hot, no one will be able to tell because overexposed snow looks like normally exposed snow, only not as white.

Use gold reflectors to help highlight stationary objects (moving objects are too difficult to light this way). A silver reflector will give a cooler, more winter-like light, while the gold looks more inviting and warmer. Following the action will also prevent objects from strobing.

Adverse effects such as lessened battery power, frostbite, condensation, and constricting clothing will make winter shooting harder, but if you ever get too hot, this is probably the right place to be. Figure 6-1 illustrates the extreme contrast range with white snow (stay away from all other colors).

Figure 6-1

Lovely white snow

#68 How to Shoot on the Beach

Sand isn't that different than snow; it's only harder, scratchier, not wet, and more abundant. Sand will get into anything and everything, but illuminating sand or recording an image on a beach is much like doing the same with snow.

The beach is an extremely bright place to be because the sand (like snow) reflects the sunlight. The f-stop on a beach is between F16 and F22. In order to cut down on the sun's incredible brilliance, it might be wise to erect a 12-foot by 12-foot silk 25 feet in the air above your talent. This will shade the area by at least four stops and provide a more even, less harsh light.

But with the strong breezes blowing off the ocean, the silk we used on a shoot immediately became a huge sail. One of the crew members instantly scrambled to grab a sandbag from the van, but remembered our location and instead decided to act as a human sandbag. While standing on the legs of our beefed-up C-stand, he was able to steady the base of the stand as the silk itself rippled violently in the 30 mph winds.

If lighting talent on the beach, use amber gels to bring up their flesh tone slightly. HMIs are daylight balanced and will give you more output than puny tungsten units. Using a gold reflector as well as several sheets of foam core may also help bounce the harsh sunlight onto your subject.

Always take an exposure reading from your subject rather than the entire area. Unless you desire backlighting, the sand and the sea are far brighter than your subject.

You probably will be using battery power for everything unless you made arrangements to get a generator. Dress comfortably, knowing you will be exposed to the elements (sun and wind). You will also attract attention with your camera, so be prepared. Lastly, keep an eye on your camera at all times because sand may be attractive, but it ceases to be when inside your camera (or shorts).

#69 How to Shoot Sporting Events

When shooting a sporting event, you must keep two things in mind. The first is to keep the players in focus as they move all over the place just so you have a more difficult time shooting them. And the second is to follow the ball. Wherever the ball is, that's where the people in the first item will be.

Let's begin with football, which some people call a contact sport. A bunch of really big guys are all chasing after an odd-shaped ball in the hopes of falling on it. These armor-clad crusaders will stop at nothing in the hopes of flattening that pigskin. I've taped high school (much safer than pro) football games where the linebackers rushing at you only weigh 250 pounds, stand at 6'4", and are 18 years old.

You need to get the best vantage point to see where the players and ball are at all times. If the football travels to the end zone, they *will* come running after it. You had better move because they will not let anyone with a camera stop them. When they finally get the ball, they give it to someone else anyway. It makes no sense to me.

My point is even though high school players are smaller than the pros, you must be there to get the shot. I've grabbed great footage just before I was knocked over, with the camera still running (like I should have been). If you see people rushing toward you and continue to keep shooting, pull your eye away from the camera. I promise you will still capture all the action, but you also will be saving yourself from a nasty injury. Surgeons don't like removing video cameras from people's nasal passages.

A much safer sport with a much smaller ball is baseball. It's almost impossible to see the ball as it is hit by a batter, so dwell more on the pitches from over the shoulder of the pitcher. Once the batter swings and hits the ball, cut to the outfield guys running. The small white ball will not be visible in the air. Once on the ground, your chances increase dramatically. But everything you shoot in a game like this is in telephoto. At least you can be on the field in football where it's far more dangerous. Nobody is going to run into you in baseball so you have to be stationed three miles away.

As the sports get slower, the balls get smaller. Golf balls are about the smallest you can shoot comfortably, but only when they are on the ground. Once again, in the air you will see nothing but clouds. Shoot from behind the golfer as he or she swings and don't worry about the ball. Focus on the ball as the golfer is putting. The green grass, the bouncing ball, and close-ups of the players will tell the story.

Outside at a sporting event like golf, it's best not to use the auto features of the camera. The auto-focus can't keep up with the tiny ball, and the auto-exposure always has the players in silhouette. In the past, every time I zoomed into a close-up of the ball rolling along the grass into the hole, it didn't. Whenever I chose a wide shot, that's when they would sink the ball.

Shoot everything as wide as possible; this is the only safe way to capture the action when you have no idea what's going to happen. Also use the camera's manual features, because the auto controls have no better idea of what's going on than you do. Don't use the flipout screen because it's impossible to see in the bright sunlight and it sucks the life from your battery. Finally, prepare yourself for exposure to the elements.

#70 How to Shoot at an Amusement Park

An amusement park is like no other place on earth. There are sugar-fueled children who take on enormous rides that delight as well as terrify. The kids are focusing solely on their amusement (thus the name), and your job may be to capture that on tape without interfering with their enjoyment.

Beforehand, know which types of things you'd like to shoot and try to get to the amusement park early to scope out the location. Which angles will work best? Does a subjective shot lend itself to a particular ride more than another? How do you capture the sights, sounds, and fun without causing harm to your body?

Try out each ride several times with your camera before you shoot anything. This will better help you determine a vantage point. On any ride, image stabilization should be switched on since your camera will shake. To avoid getting sick, don't focus on just what you're shooting; keep your other eye open so you have a reference to the horizon. Your body is moving one way and you may be seeing something else in the viewfinder.

Try to establish the ride in a long shot first so people know what you are filming. Immediately after that shot, go subjective. This is where a storyboard is helpful because each ride needs to be shot from a different perspective, and the day of the shoot is no time to be fumbling around trying to plan this.

The amusement park is open for business and you have to work around the paying customers (unless they made you pay to get in). Their entertainment comes first (not by watching you screw up) and you have to keep that in mind.

Check your equipment constantly. Movement, heat, cold, dark of night, and sun will change often, so make sure you are shooting with equipment that's functioning. Motion sickness is a definite possibility because you are getting older and you're riding more motion rides. If you think you may become ill, prevention (before the ride) is the only way to avoid it. Take legal drugs and if that isn't possible, try to focus on something else to take your mind off the sickness.

Don't change the angle too often within the same ride. You probably can't use 28 different angles of the same ride in your video. Beforehand, decide which angle works best and limit it to no more than three, which still leaves you many options in editing.

Take frequent breaks between rides to allow your stomach to catch up. This is a good time to play back the footage before you leave the particular ride. Do this discreetly because everyone will want to see it too.

Have too many spare batteries, drink plenty of fluids, and wear lots of sunscreen. You may not realize that you are dehydrating while trying to capture that elusive shot. Shoot miles of footage of everything, because not all of it will be usable due to shakiness, jitter, focus, exposure, and color balance. These items should be checked before the ride, but you won't have control over most of the ride (believe me, I lost control long ago).

Never stop the ride so you can adjust your equipment. Ride stoppage, like a movie abruptly ending after three minutes, does not win you friends and you are far outnumbered by the "little people." Check your equipment before boarding and use two or three ride attempts to set up before shooting.

If any equipment malfunctions, check the most obvious first. The on/off switch is more apt to change than a video head blowing in the middle of a take. Take your time, don't panic, and you will discover the problem before having to ask the occupant in the next seat to help you.

Solutions in this Chapter:

Chapter 7

How to Shoot Sunrises and Sunsets

#71 How to Shoot Sunrises and Sunsets: When to Underexpose

One of the most beautiful images you can record on tape has to be a sunrise or sunset. As the large, orange sphere rises or descends into the horizon, it's your job to capture and record the colors of this illusive target.

Framing isn't too much of an issue, but try to get a foreground object in the shot. This will give your viewer more perspective and will better help you determine the exposure.

Because of the brightness of the sun and the surrounding sky, you must determine if you would rather under- or overexpose slightly. The key here is *slightly*; video does not handle extremes too well.

If you want to expose for the sun and sky and have all other objects in silhouette, underexpose your image. Take an exposure reading of the sun and sky and set your camera's iris accordingly. You will have no detail in anything else (very black) but the sunrise/sunset, and the viewer's attention will be called to the orb in the sky with its many colors. Because of color temperature, the camera will see hues of orange and blue that we can't normally see (unless we are cats, because dogs see in black and white—how do we know that?).

The further you underexpose, the darker the blacks will be and the more color will be seen in the sky. The sun itself is eight stops brighter than the sky, but people are more interested in the colors and just use the sun itself as a visual reference. It would be impossible for the audience to tell whether the shot is of a sunrise or a sunset unless you specify the direction in which you shot. Figure 7-1 illustrates an underexposed sunrise.

Figure 7-1

A magnificent underexposed sunrise in glorious black and white

#72

Figure 7-2

Exposing for the foreground—this is what overexposure looks like. (Photo courtesy of Linda Gloman)

#72 How to Shoot Sunrises and Sunsets: When to Overexpose

If you want to see detail in the foreground objects (people or scenery), you need to overexpose. Your Aunt Martha standing under the palm tree will look better (if that's possible) if you expose for her rather than taking a reading off the sun. The colors of the sky will be more muted and washed out, but the world will see your relative or the scenery.

Once again, you must be subtle in your overexposure because video cameras don't like extremes in anything. Pushing the backlight button on a consumer camcorder will achieve this same thing automatically.

So if you want to see the sun and sky, underexpose. If the person or object holds more value than the sky and sun, overexpose. I surprised you in that I didn't make any jokes about exposing yourself; I actually am maturing. Figure 7-2 shows an example of an overexposed sunrise.

Chapter 8

How to Shoot in a Studio

#73 How to Shoot Effectively in a Studio

One of the most controlled environments on earth is the studio. You are sheltered from the elements and you have access to lights, power, sets, and sometimes equipment. But how can you make all that work for you?

In this chapter we'll discuss using China balls to cast an even light on things, how to choose the right diffusion material as well as testing how each type works, using chroma key to put your talent in front of someplace they really aren't, how to assemble and light a set from a kit, and how *neutral density* (ND) gels can be your best friend on a set.

#74 How to Use China Balls

Although not something you serve at Thanksgiving, China balls (an enclosed paper shaded light) will produce a lighting quality like no other and can be purchased at Ikea for less than $4 (minus the photoflood bulb).

Our grip, Brad Kenyon, used China balls as our fill lighting for a flooring shoot. China balls or Chinese lanterns are manufactured by Chimera, Lowel, and a number of other sources, but our low-budget specials came from Ikea. The only real difference between our lights and the professionals were that ours had open bottoms and weren't completely enclosed; this created one drawback.

In the cutting room where all the laminate flooring was cut to size, we used the Chinese lantern as our only fill light. The floor-to-ceiling picture window offered gallons of daylight, so we opted not to gel the window and balance the camera for daylight.

Figure 8-1

Using China balls on a set

#74

The installer would have plenty of natural light without the heat created by tungsten lights.

In order to soften the harsh shadows created by the ungelled window, we used a Chinese lantern with a daylight-balanced (blue) photoflood bulb. If we had decided to go the tungsten route, a standard 3200K photo bulb would have replaced the blue bulb. This 300-watt unit was hung from an extended C-stand arm over the installer's cutting area. The 360 degrees of soft light filled in all the shadows and gave us a very pleasant look. Figure 8-1 shows Brad and his nifty invention.

However, since the lantern was an inexpensive unit, the photoflood's light shown through the hole in the bottom of the ball and cast a circular ring on the work surface. This unnatural source was extremely distracting and was soon covered with gaffer tape. A piece of diffusion would have also softened the problem and given us light, but in our case we wanted no light and black gaffer tape was the best solution. The light was still allowed to vent out of the top of the unit. If not, we would have soon had a very flammable light source.

In every shot for the rest of the installation, the Chinese lantern was used. In the dining area, the 300-watt daylight-balanced photoflood was swapped for two 150-watt tungsten-balanced photofloods. The two lanterns provided the same type of fill as the single unit with twice the fun.

#75 How to Choose the Right Type of Diffusion

People go out everyday and spend good money purchasing lighting equipment, and what do they do? They cover these nice bright lights with gels and diffusion material. But how do you decide what's the best type of diffusion to use in a particular case?

When you shine a light on someone or something (a three-dimensional object), shadows are created. These shadows exist because no light diffusion is present (or past or future for that matter). Look at the biggest light source around: the sun. On a cloudless day, shadows are cast all over the place because nothing is stopping or diffusing the rays of the sun. When the clouds roll in, we have diffusion and the light is much softer. Mr. Science says water vapor in the clouds spreads out the harsh sun's rays, much the same way diffusion disperses the rays of a lighting instrument, only without the water vapor (a lot less shocking too).

The biggest difference between the sun and our lights is that the diffusion (not the clouds) we place in front of the lights is controllable and we get predictable results. Don't even try to get controllable results with a cloud; there's no talking to them. But do we really know how predictable?

By doing a test, you can determine which type of diffusion works best for you. In conducting this test or any test, you must have a set of nonvariables or things that will not change between each image to determine what works best. In my test, I chose a model and had her wear the same clothing test after test. I had variations in only the diffusion and noted the differences. It is imperative that you document by photos and notes so you can see and read the subtle variations.

For the test, you need a still or video camera to note the visual changes in the lighting and shadows. Whether you use a digital camcorder or an analog, 35mm, or digital still camera is not important. What *is* important is not to change your recording method during the test. You need a standard to base everything on, so use that standard in all your tests.

A small side note must be made if you are using 35mm still film. If you are shooting indoors under tungsten or HMI light (daylight), use slide film rather than negative. When slide film is processed, you still have a reversal or positive image. The computers at processing labs don't try to compensate for slight color shifts, color temperature, or exposure like they do with negative film. To reuse that tired old phrase, with slide film what you see is what you get. This is a much better way to control your test.

With a digital still camera, choose the correct setting for color temperature and don't make any variations to the image in Photoshop or any other enhancement program. The same definitely applies if using a video camera.

#76 How to Test Different Types of Diffusion

Have your test subject sit or stand in the same position for all the lighting and diffusion changes. It may sound silly, but make sure your model does not smile between the images. Sometimes you might like the look of an image better because the model smiled rather than the change in the lighting.

Once again, keep everything consistent. I'll explain how I test diffusion and you can make variations to suit your needs.

On a particular shoot, the model sat in a backless office stool 24 inches from a gray paper backdrop. A 600-watt Omni light with a chocolate gel provided a slight illumination streak to the background and gave her some separation (no, she's not divorced yet). In addition, a 100-watt Pepper light was stationed camera right, 3 feet above her, and acted as a kicker on her blonde hair. These lights and background stayed constant through all the images.

To determine which silk worked best, no fill lighting was used to detract from the look. The key light throughout my test was a 1K Fresnel. As the diffusion on the Fresnel became heavier, I moved the light closer in order to achieve the same f-stop (remember, I wanted to be consistent). All my images were exposed at F5.6 with a 100mm lens. I chose a telephoto lens because I wanted to see minute detail and the nuances some of my different setups would create. A higher-quality lens and a better camera will help you decipher smaller, more miniscule changes.

A million different kinds of diffusion exist out there and I had to draw the line someplace. Get a sheet of each type of diffusion you may like to use and begin your test.

Test Shot #1, *no diffusion*: You need a shot like this to see what an undiffused Fresnel will do. The light is only slightly modeling because the beam is going through the glass lens in the front.

Test Shot #2, *Tough Spun*: In Rosco's book, this is considered slight diffusion and gobbles up two and a half stops of light. I noted a slightly different appearance in the highlights of her face. You can actually tell some type of diffusion is in front of the light.

Test Shot #3, *Tough White diffusion*: This type of diffusion is considered moderate and uses three and a half stops of light. In our shoot, the model's face was more evenly diffused with the almost invisible thin shadow lines the Tough Spun created.

Test Shot #4, *Opal Tough Frost*: This piece also belongs to the slight family and takes a modest one stop. It shows that within the slight family (sounds like a *Saturday Night Live* skit) there is a diffusion that absorbs very little light.

Test Shot #5, *Soft Frost*: This one's in the moderate group and takes twice the amount of light as our last contender, two stops. I noted that a soft prefix in front of the diffusion gave a mistier look, while the opal changed the skin tone slightly. The word tough, though preferable in bars, just means the diffusion material is more durable.

Test Shot #6, *light grid cloth*: Welcome to the heavy diffusion family. This diffusion is so macho that it even comes with its own metal frame. That's probably why it takes three and a half stops of light and has a drinking problem. This diffusion was readily noticeable on the model's face. This actually gave her a "painted" look like viewing an oil painting.

Test Shot #7, *net*: Just for the fun of it, I wanted to climb up very high without using a net, but for this test I used one. Nets are also in the heavy grouping (you have to kiss their rings) and used two and a half stops. The look was much the same as light grid cloth.

Test Shot #8, *foam core*: Just to live on the edge, I bounced the Fresnel into foam core and got a drastically different look. This bouncing absorbed two stops of light but produced an even illumination.

Test Shot #9, *silver umbrella*: The umbrella reflected my Fresnel rather than bounced it. Only one half stop of light was lost in the process and the silver umbrella did nothing more than raise the color temperature slightly towards the blue spectrum.

I tried the same tests with a 600-watt Omni light and the results were quite different (between the lighting instruments, not the diffusion). In these cases, unlike filters, looking through the diffusion material does not show you what the subject will look like. A lighting instrument must be used to give you an accurate portrayal of the result.

If you keep things consistent between setups, you will have an accurate test to better help you determine which silk works best for you. Table 8-1 gives a breakdown of each silk and how it performed.

Table 8-1 Diffusion Description

Test Shot Number	Diffusion Material	Light Loss in Stops	Family	Results
Image #1	No diffusion	N/A	N/A	N/A
Image #2	Tough Spun	2 1/2	Slight	Fine shadowed areas
Image #3	Tough White	3 1/2	Moderate	Softer, more even
Image #4	Opal Tough Frost	1	Slight	Changes flesh tones
Image #5	Soft Frost	2	Moderate	Mistier look
Image #6	Light grid cloth	3 1/2	Heavy	Oil painting
Image #7	Net	2 1/2	Heavy	Oil painting
Image #8	Foam core	2	Moderate	Flat, even
Image #9	Silver umbrella	1/2	Slight	Cooler

#77 How to Use Chroma Key Effectively

George Winchell, our lighting guru, created the perfect chroma key lighting setup. The main thing (or should I say key) to remember about chroma key is to light the blue (or green) background as evenly as possible.

Any shadows, light spots, or wrinkles might not key out as well in editing, and that would definitely give your secret away. Since we were shooting an interview (actors in front of the blue screen) and adding the backgrounds in postproduction, we did several tests with a light meter to make sure the background was evenly lit. Walk the entire length of the backdrop holding the light meter. Any high or low spots (uneven lighting) should be corrected.

The object is to "drop out" or "key out" everything that is this certain color. All blue or green in the chroma key background will be replaced with something else (make it something pretty).

Using two open-faced lights (2000 watts), one on the left and the other on the right, we had an even F4 background. Use your strongest lights to accomplish this; they may be open-faced or Fresnel. In our case, we diffused the open-faced lights with Tough Spun to soften the light on the blue surface. The only purpose of these two softlights is to illuminate the background. Pointed at a 45-degree angle to the background, each light was 6 feet high and pointed at the screen. By moving each light carefully and playing with the barn doors (they may be opened or closed to block light), we had an evenly lit sea of blue. The talent needed lights of their own. The talent should also be a few feet away from the chroma key background; you

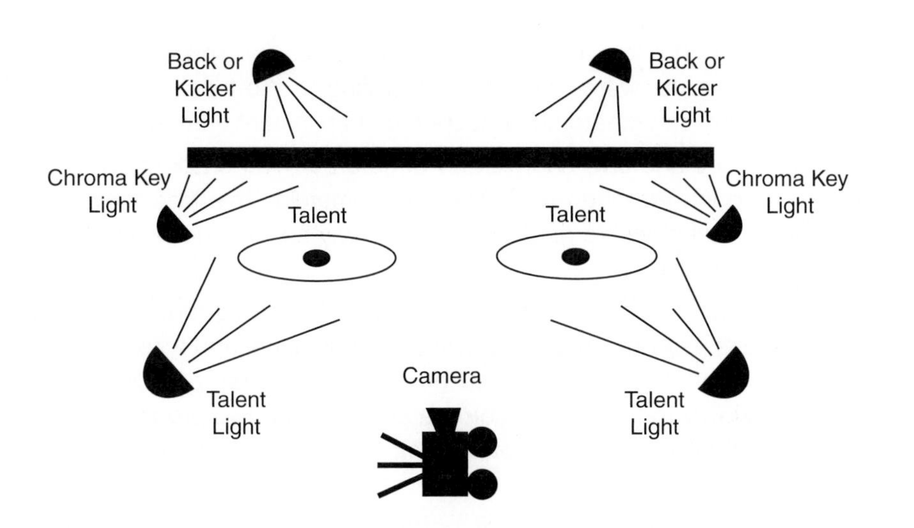

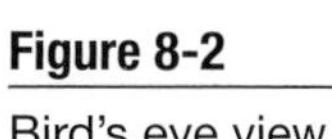

Figure 8-2

Bird's eye view of chroma key setup

want their shadows to fall on the ground rather than the backdrop. Figure 8-2 shows the correct positioning of lights and actors in relation to the background.

Some people prefer gelling the lights on the chroma key background with yellow. This color accentuates the blue and makes it even harder to duplicate in the real world. This isn't necessary but might make your task easier.

We used softlights (Lowel 2K) to light the talent. Each one of these softlights was used as a key light for the talent, one for the interviewer and the other for the interviewee. Once diffused even more with Tough Frost, the talent had an even, soft highlight on their faces. In fact, some of the interviewer's key light acted as fill light for the interviewee and vice versa (refer to Figure 8-2). The softlights were placed high enough so the soft shadows they cast fell on the floor rather than the blue background.

The talent's backlight or kicker is even more important in a chroma key shoot. They must be separated (or have dimension) from the background. Remember, in love, war, and chroma key, everything must be even.

By using a net in front of the softlight, the talent's face would appear less washed out and have more detail. You're free to light and model your talent any way you like using conventional lighting; the only trick is to keep the background free from shadows. This shoot could have been lit like any other interview setup; we just had to make considerations for our chroma key (and you should too).

#78 How to Create and Light a Set from Scratch

Every once in a while you are asked to create something that isn't there. Without resorting to black magic, you can build and light a set that didn't exist previously. The key to this magic act is portability. This is the reason location lighting kits were created. By traveling to the location with your lighting kits, you can create an atmosphere of an instant television studio.

A client called and said she wanted to do a 20-minute talk show about her new product. Although more like an infomercial, this talk show host would ask knowledgeable guests to explain why her product worked better than anyone else's.

We were given permission to shoot in an 800-square-foot room. Two hundred chairs had been set up for a viewing audience, and a set was created in the front. The set consisted of three 8-foot gray cubicle panels, flanked by two 10-foot pieces of corrugated Plexiglas. Two prop plants and three chairs filled out the front of the set. Figure 8-3 illustrates our work of art.

All of our lighting consisted of portable instruments that had been brought to the location in a lighting kit. AC power would be obtained from outlets in the floor.

The first thing to light would be the two Plexiglas panels on the left and right of the cubicle partitions. The client wanted each of these panels to appear blue. Placing lights behind the panels and applying blue gel solved this easily. A 1000-watt light, gelled with color correction blue, was placed behind each panel. At a height of eight feet and at a distance of six feet from the Plexiglas, each panel now had a soft, blue cast.

Figure 8-3

A set created from nothing but a kit

The prop plants on the left and right of the set had to stand out from the blue Plexiglas panels. A 750-watt Lowel Omni was placed on each knee wall and pointed at the plant. The diffusion of choice was also 216. The Lowel Omni Kit acted as our plant lights.

The light on the left, extended to a height of 7 feet and a distance of 10 feet, acted as the key. The other DP, also at a height of 7 feet but a little farther back at 15 feet, acted as the fill. The talent now would look less like typical talk show guests with a slightly lower key lighting.

The talent's backlight was tackled slightly differently. A 1000-watt light with 216 was attached to a C-stand arm and extended over the tops of the gray cubical walls. This 1000-watt light at full flood would act as the backlight for all three people on the set. We now had used all four lights in our Lowel DP kit.

Once the cables on the set were dressed, the audience was invited in. It's important to gaffer tape anything that comes in contact with people's feet. If it can be stepped on, stepped over, or pulled out, tape it down.

At the end of the day, the set could be torn down and folded up in an instant. All lights would be returned to their respective cases and no one would ever know that we had shot a talk show in that large space, thanks to location lighting kits.

#79 How to Use Neutral Density (ND) Gels

Neutral density (ND) gels are worth their weight in gold. Actually, they're worth much more than that because they weigh relatively little. I had to light a white ceiling with installers and not let them blow out exposure-wise (overexpose them).

If I diffused a 2K with Tough Spun or plain old 216, the lighting on the installer's face would look fine, but the white ceiling would always be two stops overexposed. In video, anything two or more stops overexposed is grounds for stoning.

Since the white ceiling tiles were exactly two stops over, attaching a ND6 gel to the top of our diffusion allowed the ceiling to return to a slightly more normal pallor. The ND gel would absorb just the right amount of light to keep the ceiling and our installers at the same f-stop (I didn't want them to fight). If the ceiling were more overexposed than two stops, we would have used an ND9 and an ND3 if the tiles were less bright (not stupid, just not as smart).

Using these types of gels as a partial light blocker or as a transition or gradient light blocker kept things on an even keel. Of course, we could have used our ND over the entire piece of diffusion, but it would have cut the light

Figure 8-4

Our diffusion/ND setup for the set

too much where we didn't really want it cut. As Figure 8-4 illustrates, all our lighting would have to pass through this mixture of diffusion and top third neutral density. Every time we set up a light, this combination ND6/216 gel/diffusion would be set up in front of the unit. Since we had to have the ceiling in every shot, we had to be consistent with our exposure, and this combination did the trick.

ND gels, or filters for that matter, equally reduce the amount of all visible light at the wavelengths at which they are transmitted. It reduces the quality of the light (amount) and not the spectral energy distribution. If you're really proficient and always shoot at a particular f-stop, a ND gel or filter can help you achieve that elusive number.

If we needed a little extra punch in our lighting setup, we'd attach a 1000-watt light to the lower portion of our 2K light and stand. Half the 2K's light would still be shining through the ND6/216, but the 1000 watt light would just be shining through the 216 diffusion.

If I have to light something and a window is in the shot, this is where the color correction ND gels come in handy. I've rarely used just the 85 gels alone on a window. The 85 gels will correct my color temperature from daylight to tungsten, but outdoors is usually at least two or three stops brighter than what I'm shooting indoors. Here is where the 85ND6 (two stops) or 85ND9 (three stops) comes into play. The color temperature is corrected and the light level is lowered. Everybody is happy in this case, except the poor souls that have to gel every window in the shot.

Chapter 9

Specialized Production Challenges: *The Creation Process*

#80 How to Create Something from Nothing

This chapter discusses creating things that aren't normally there. Using your ingenuity, limited budget, and the resources available, you can learn to fool the viewer into believing something that isn't so.

With little time and money, you can accomplish the following:

- Build a cinder block wall that can be torn down and re-assembled at will with no mess
- Create fake sunlight streaming through a phony window
- Light a dark coat that absorbs light
- Make an animal talk to the viewer by moving its lips
- Deceive the viewer into seeing something
- Create realistic fire on camera
- Make a small amount of anything look like much more
- Make two different cameras look like the same one
- Teach an actor to react to something that isn't present and will be added later
- Use subjective shots effectively to pull the viewer into the action
- How to show a TV set displaying a picture

#81 How to Build Disposable Cinder Block Walls

Did you ever have to build a semi-elaborate set only to have to tear it down because you no longer had use for it? There's a quick way to do this and I learned it from my friends at Megcomm Video Productions (www.megcomm.com).

Their client needed to see a cinder block wall set, one that the camera could dolly around from the front side to the back, basically a cutaway wall. But the masons out there (the people who work with cement and mortar, not the Masons who belong to a club) know that once you put mortar between cinder blocks, the wall becomes a permanent fixture. This is probably why people use mortar to build their houses. No one wanted their walls to be cutaway for this shoot, so something else had to be done.

Once this cutaway wall was constructed and videotaped, it would have to be dismantled with a sledgehammer. Mike Gorga and Joe D'Angelo from Megcomm had a better idea that was faster and cheaper, and allowed them to use the cinder blocks over again.

This time Martha Stewart came to the rescue. Instead of using mortar between each cinder block, Joe purchased gray Martha Stewart towels from Kmart. These towels were folded in between and beneath the cinder blocks. With a little time spent folding and removing any creases, the towels looked just like mortar and didn't have to dry.

The human eye and camera could not differentiate between the towels and real mortar. The viewers (and client) were once again fooled by an inexpensive device that allowed Megcomm to remove the wall after the shoot. The next time you use a towel, think of what else it may be used for. Figure 9-1 shows how realistic this looks.

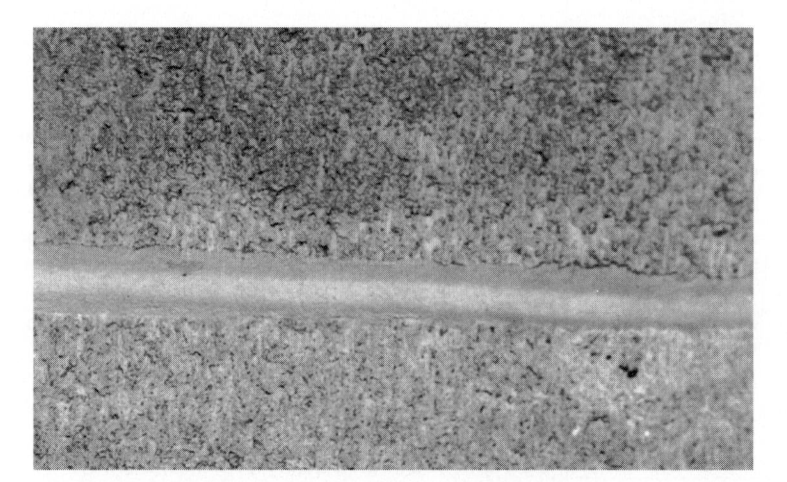

Figure 9-1

Cinder block walls with mortar towels

#82 How to Fake Streaming Sunlight

Sometimes you want the look of sunlight streaming through some windows to occur where you have no windows. Your only option is to create that effect with the tools you have on hand.

Although the sun tried to stream through our window, the gel we used had softened and absorbed most of the light. Any sunlight falling on the floor had to be created by us. As the clouds blackened overhead and the storm approached, we hurried to create our effect.

All modern homes have electrical outlets outside, which are very convenient if you want to plug lights in and not run the cords through an open door. We tried using our ungelled 1000-watt DP as our sun. We left the tungsten light bare because the 85ND9 gel on the window would convert the color temperature. Unfortunately, one DP isn't enough; neither are two. With 2000 watts of light pounding through the gel, no discernible pattern appeared on the wall or floor. The gel was taking all light and diffusing it too greatly. The only options were to not gel that portion of the window or bring our sun effect inside.

We tried to shine the light through the ungelled portion of the window. The double pane of glass still softened the image too much. Once the light was gelled with booster blue (to give it a slightly orange hue), we lost all definition on the wall. We decided to bring all lighting effects indoors where we would have more control.

Tom Landis, our lighting expert, first tackled the slash or orange light falling on the wall (from the sun). A 750-watt Omni was gelled with amber gel (to make the orange sun streak more pronounced), and a screen was inserted into the Omni's barn doors to cut the light's output. In front of the gel, Tom cut horizontal slats in a piece of black wrap. As the light was placed close to the wall and very low, the orange streaks on the wall looked like the evening sun filtering through the window. Figure 9-2 depicts the streaking device.

Tom then found an empty picture frame in the model home. He created a window pattern with black strips of gaffer tape and attached them to the frame. Orange gel was placed against the sticky side of the tape to create the orange light he was after. Essentially, he had created a window that he could hang in front of any light.

With the pseudo-window attached to a C-stand arm and knuckle, a 1000-watt DP and a 750-watt Omni acted as our key source. Angled down, the orange streaks cascaded over the floor and still showed the window panes. Isn't it amazing what $2 will do in the right hands? I would have spent it on candy. From now on when I want to see sunlight streaming through the windows, I attach gaffer tape and orange gel to my glasses.

Figure 9-2

The only way to streak legally

#83

#83 How to Light Dark, Furry Things

If you've gotten past the title of this section, you probably know that I'm talking about minks, sables, chinchillas, gerbils, hamsters, mice, or foxes. You probably light all these animals the same way, but I'm really talking about coats—fur coats. I really don't wear fur coats, and my raccoon coat is still at the cleaners. Besides, I think my beard contrasts too sharply with furry coats. But that story is a little hairy.

A furrier brought his furs to our studio and we had employees and friends model them. He had more money invested in his rolling cart of pelts than most people make in a lifetime. That's when I found out I was allergic to chinchilla. You're not supposed to sneeze on a $15,000 coat.

I'm told that it takes quite a few mink or chinchilla to make a coat. I thought that each mink just unzipped their fur and stepped gently out of it. That's the way they did it in the cartoons. You can now see how impressionable I am.

Once we set up the camera and monitor, every coat in the studio was pitch black. Actually, they all were varying shades of black, but still black. These furs are light suckers. Every light in the room was pulled into this black hole of the fur world. Even the white coats looked a little peaked. It seemed that these animals absorbed light. They had sucked all the light out of the studio and were working on the place next store.

We quickly set up our Lowel Omnis and pounded them into the coat. The light actually helped; it really brought the texture of the fur to life (poor choice of words). These black, lifeless orbs now could be seen as fur, but each coat was absorbing 750 watts of light. However, the chinchillas still looked as if they had five o'clock shadow instead of being chocolate brown and inviting. With the addition of an amber gel on the 750-watt Omni, the coat looked great. In real life, it looked as if we had 750 watts of lights shining on a coat. But through the monitor, the coat absorbed three stops of light. It was perfectly exposed. The darker coats needed no help from the gels.

We traveled to the fur dealer's store to capture wide shots of his inventory. As if on cue, a businesswoman and her college-aged daughter walked into the store. Seeing me standing there next to the camera, I guess she though that I might have known what I was doing. She said she wanted to see a sable. Tired of always being confused with Kmart employees, I told her the Mercury dealership was down the street. Finding very little humor in my statement, the large woman huffed and walked over to the rack.

I asked the woman's daughter if she might be interested in modeling one of the furs for us. Without speaking, she threw her coat on the floor and stepped over to our lit rack. She put on a fur and immediate turned into a supermodel. It was as if she had been doing it for years. Without missing a beat, the crew pointed a light at her coat, set another Omni up as a kicker, and a third bounced into an umbrella. The girl found a nearby mirror and began spinning.

The client, from a distance, came running over and began pulling other furs from the rack. He pulled a black sable that cost more than my house. "Here, wear this," he said as she donned the garment. With one hand under his chin and the other stroking the hair on his arm, he had made the discovery of the year. Figure 9-3 illustrates a beautiful object (the mink coat).

I guess you could say that the key to this whole experience was to be ready for anything. My advice on a low-budget shoot like this is to always be ready with a couple of lights and gels; you'll never know when your best model may show up.

#84 How to Make Animals Talk

Of course, the animals can't really speak, but the illusion of them moving their lips and words emanating from within is the concept. You all loved when the animals in the *Babe* films spoke, but how can you accomplish that with your equipment?

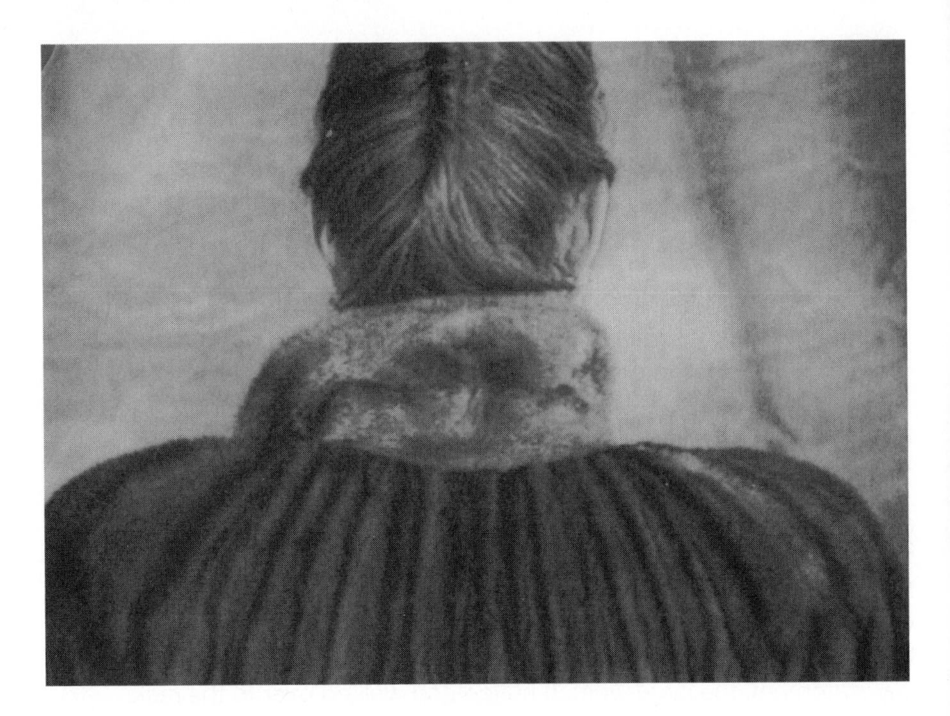

Figure 9-3

A nicely lit mink coat

I know that some would say it is easy enough to create that: It happens every evening on prime time with talking dogs, cows, cats, and a lizard. It has almost become common place to see some sort of animal talking on TV. We've come a long way from *Mr. Ed*. Two methods exist for creating lip movement in animals: the cheap, sloppy way and the digital way.

When I edited a spot that required talking animals, I did it on an analog system. The preferred cleaner method is with a digital unit, but I'll explain the cheap route first. I had to make the pets talk the old fashioned way. The audio track was easy. I just laid down the stereo mix of the jingle. Since the animals would be singing the words to the song, I would have each character (pet) supply a different vocal.

Using the dynamic tracking feature on our deck, I jogged the tape back and forth to make the pets' mouths open and close like a stutter. In much the same way a rap artist scratches a record back and forth to create sounds, I did the same thing with video. I know that it sounds archaic, but jogging the images back and forth while recording onto another tape gave me very realistic effects. By looking in the mirror and watching my own mouth move, I tried to match the puppy's mouth to my movements with the words of the jingle. Of course, I wouldn't get the same fine movements that a human mouth and lips allow, but a rudimentary open/close, open/close movement proved to be quite effective. It was very time consuming, however, to manually create this same open/close sequence to match the words of the jingle.

Once the jogged video footage had been recorded to another tape, I began to apply it to the submaster tape with the jingle. By adding or subtracting one frame at a time, I was able to get the open/close sequences extremely convincing. At times, I had to use a two- or three-frame dissolve to join the open/close jaw sequences so the viewer wouldn't be seeing a jump cut. At other times, I had to create a soft-edged circle wipe directly over the animal's jaw using different footage of the pet itself. Much the same way Kennedy's speaking was accomplished in *Forrest Gump* (only without the digital effects), I had to mix a moving mouth onto a nonmoving character. At least my effect didn't look as cartoon-like.

At various times throughout the jingle, dogs barked, parrots squawked, and cats meowed. When these sound effects were heard, I would cut from the animal talking to a dog barking and sync it to the audio. I did the same thing when the audio had a parrot squawking. Throughout the spot, animals were singing the words of the jingle, dogs were barking, and other pets were just generally joining in the fun. It's hard to believe that all that jogging work would be over in 30 seconds of screen time.

The digital solution to talking animals only takes a little longer than jogging the footage because of rendering. The effect is created by getting the footage of the animal into your system and isolating the mouth area. This can be done in two ways.

The first approach is to use a software program like Illustrator, Photoshop, or a number of others out there. Map the mouth area and then erase and clone sections until you have the open mouth you desire.

The second approach takes more artistic skill and involves animating the mouth positions. This is the most realistic and will be explained in the software's manual. Because of the tight schedule, budget, and someone else using the digital system, the quick and dirty approach worked for me.

But we have a choice and that's why we are smarter than the animals, at least the ones I recently worked with.

#85 How to Deceive the Viewer

Sometimes you will want to create a specific look that isn't exactly what the camera sees. How do you change that, or mildly deceive the viewer to show them only want you want them to believe?

Not wanting really to lie, I just had the intention of stretching the truth slightly. The first shot in my video opened with an out-of-shape woman weighing herself on one of those doctor-type scales with the sliding weights. Since we wanted to give the viewer the impression that she was really out of shape, we desired to exaggerate the fact that she was extremely heavy. The actress doused herself with water (everyone knows you always weigh less when you are dripping wet after taking a shower). The only difference was that this woman was wearing six pairs of sweat clothes. The multiple layers of clothing made her slim body sag and droop in the right places. I didn't want to offend anyone by actually using someone that was heavy, so instead I chose an actress that was in shape and made her look as if she was in need of a work out (politically correct). Figure 9-4 showcases our talent on the scale.

In order to make the woman appear heavier than she really was, we placed two 60-pound flat barbell weights on the bottom of the scale and had her stand on top of the weights. As she slid the lower weight across from 50 to 100 to 150 to 200, the scale wouldn't balance. It was obvious that she wasn't really that heavy, but to add to the humor of the scene, we had her eating a Twinkie while trying to weigh in. As mentioned earlier in this book, the best approach in humor is to be subtle; don't hit someone over the head with your joke. You actually will get a bigger laugh with something that is subtle rather than obvious. The key is to build upon this humor so one joke immediately follows another. Some of the jokes may be missed,

Figure 9-4

Six pairs of sweats and she still has room for more.

but it's better than spending all your time and money on a big gag and having no one appreciate it.

In order to further enhance the fact that she was out of shape, we devised a series of scenes that would have her appear to be deep in her couch potato mode. For instance, the next shot involved our talent lying on a sofa, surrounded by food, watching an exercise video. The room was made to look like we had taped the scene in the late evening. All the shades were drawn, the camera was underexposed two stops and three of our 600 lights were gelled with full booster blue. This gave us the bluish, nighttime effect we were looking for. Since the other three lights were blue gelled and pointed directly at our talent, everything had the blue, harsh, directional look we wanted.

Every audio effect in the shot was amplified. Her chewing was Foley enhanced (more like the sound of a cow grazing), the exercise tape that was playing on our fake TV was a woman counting (that's what usually happens in aerobics classes), and the rattling of her Cheetos in the bag as she ate them was enhanced to sound like a small brush fire. Throughout the scene, the actress was doing arm curls with a can of soup. Although the scene only lasted three seconds, we had a lot going on visually.

We were able to borrow some antique exercise equipment from a used furniture store. The first item we obtained was a fanny exerciser/shaker. When we got the exerciser to our set and plugged it in, the motor buzzed, but the leather strap would not move. We had the actress connect both ends of the leather strap to the bolts that extended from the machine and wrap the back of the leather strap around her derriere. In order to make the machine look as if it were operating correctly and make her lose weight, we had a grip lie on his stomach and shake the exerciser as violently as he could. With the sound of the buzzing and the actress being shaken and not stirred, we achieved the effect of an antique weight loss program.

I wanted the rest of the white-walled set to look stark, so a 600-watt light with a Tough Spun silk covering was our key source of illumination. The silk in front of the light created a soft, larger-than-life shadow on the wall of our actress exercising. Another light with a bastard gel provided the kicker light to separate her from the white walls. As she ate potato chips, shards of chips would fly out of the bag and crash to the floor. The grip soon complained that he was getting a better workout shaking the machine than our out-of-shape woman was getting.

We were also able to obtain an antique exercise bike from the same furniture store. While pedaling the bike, the handlebars can be pulled toward or pushed away from you. To make her exercising more exciting, we attached a chocolate donut via a fishing pole and line to provide inspiration for our talent to exercise. I hadn't realized how sharp the fishing line was, because each time we were ready to shoot, the line would slice through the donut and make it unusable. At least the mice would have plenty to eat. We also created a mock venetian blind effect on the white walls with slits cut in poster board and an orange-gelled light shining through the makeshift cookie.

The grand finale involved our actress trying to do a chin-up with a metal closet bar. When she tried to pull herself up, the bar would snap in half, sending her crashing to the ground. The metal bar was actually a one-inch diameter wooden dowel purchased from a home improvement store. The dowel was sprayed with chrome paint, and on camera no one could tell that it was made out of wood. The dowel was attached to two C-stands between the doorjamb. As the dowel rested in clamps, I sawed halfway through the wood so that it would break more easily. Our actress was lit from below to make the shadow behind her appear huge. This gave the same effect as someone holding a flashlight under his chin, not very flattering but quite effective. In addition, when editing I removed the color from the shot to make it appear as if it was from an old black and white horror film.

After several attempts, the bar finally snapped as she pulled herself up. But when the dowel broke, it also slid down the door's molding, leaving a silver streak of chrome paint at least two feet long. The snap sounded as if a mighty oak tree was collapsing in the forest. Also in editing, I kept the original snapping sound and added a CD effect of a train crash to enhance the sound of the dowel cracking.

Deceiving is quite easy if you use the correct tools. I just hope I didn't deceive you into believing I knew what I was talking about.

#86 How to Create Fire on Camera

Unless you're sadistic or have a death wish, fire can easily be recreated on video without having to leave the location smelling like burnt toast.

In the old days when people shot in black and white, fire did not show up clearly. You could see things burning, but the actual flames were (and still are) very difficult to see because they are translucent. In old movies, you rarely saw the flames, but always a lot of smoke.

Without blowing smoke at you, that is the most identifiable thing about fire. With smoke lingering around, people assume flames are also present without needing to see them. Special-effect fires are harmless, but I don't believe they look real on video. Filmed movies are another story, but on tape the image of fire is at least three stops higher than anything else and will tear and bloom when shot on a less expensive camera.

Keeping safety first, it's better to show some images of isolated flames and smoke (real frames, fake smoke) and use the magic of editing to put your talent in the scene. In a controlled area, videotape the flames from various angles. Fire itself rarely produces much smoke (because oxygen is present); when a fire starts to die, that's when smoke really appears. If you want both a blazing fire and smoke at the same time, you need to add the smoke from either a smoke cookie or a fog machine.

Both of these types of smoke are very white in color, whereas real smoke has a gray cast. Once your dummy smoke is generated, waft it into thin clouds. The effect of less smoke is more realistic than too much phony white smoke. These wisps drifting by with the fire will do the trick.

In order to show your talent at the scene of the fire, simulate the burning and smoke effect in a close-up. Once again, the magic orange gel should be used to get the color of flames on the talent's face (not too close because flames aren't *that* orange). Since flames move, I like to crinkle the orange gel to get the movement as well as the flickering color. You can also make friends with your sound person by the same crinkling noise. This way he or she won't have to record the sound of flames (they actually are mute and don't make any sound—flames, not sound people).

Experimentation is the key, so change the crinkling rate, the light distance from the subject, and the amount of smoke to get the look you desire.

#87 How to Make a Little Look Like a Lot

The first thing we will look at is how to make the viewer believe more of something exists than is actually there. Everyone's favorite object to do this with is money. How do you make a little bit of money look like a lot? The easy way is to put the denomination of bills you want on the outside of the pack and fill the center with paper slips. This isn't a new concept; just talk to someone held for ransom.

Bags of money can be filled with anything; the bag just has to look like it's full of money. In a recent commercial, I had to shoot a money tree. An actual tree was brought onto the set and new $100 bills were folded end over end to make a fan shape. The lighting used was simple but effective. The tree was placed on a table covered with a backdrop. The same backdrop was used behind the tree. A 2000-watt Fresnel was aimed down toward the tree to backlight its money-laden branches. The feathery edges of the tree would glow slightly from the tungsten light. Another 2000-watt Fresnel was bounced against a piece of foam core acting as the key light. In this setup, no fill light was used.

When shooting money in a tree, a more directional approach with lighting works best. An extremely strong backlight (2000 watts) and a bounced key light are all you need for illumination. The white and green paper money now stand out from the evergreen surroundings. A light reading should always be taken from the front, even if your backlight is as strong a source as ours was. We wanted the money to radiate wealth from within the tree, and strong backlighting makes this possible.

Deception is the name of the game and this phony illusion of depth will allow the extra money to go toward a better cause.

#88 How to Make Two Cameras' Images Look Alike

When using two cameras on a shoot, how do you get the images on both to look comparable, especially if you have two different makes of cameras?

Balancing cameras tied to a switcher or each recording to their own decks requires the images to look similar in hue, saturation, phase, and luminance. With the cameras set to color bars, begin by adjusting the horizontal phase (this is accessed by opening a small door on the side of the camera and rotating the Potentiometers (pots) with a screwdriver). With both cameras' images on the monitor, wipe half of one camera on the top of the screen and the other on the bottom. Adjust each camera so the edges of the color bars line up. If you have more than two cameras, repeat this adjustment until all are in phase.

The subcarrier phase aligns the color between cameras. With the color bars still displayed, adjust the cameras until their images match on the monitor. If the switcher is set to mix, dissolve between the sources. A vectorscope is extremely helpful to establish a norm for the color balance, but the important thing is to have the cameras looking alike.

With the lighting tweaked and ready, we powered up both cameras on our shoot. The three-shot and left-shot close-up camera was a Sony. The lens was a Canon 12:1 zoom. The host and right-shot close-up camera was an Ikegami with a Canon 6-48mm zoom. The Ikegami was cabled to a Sony Betacam SP deck. Both cameras, once white balanced, looked totally different.

In order to match the image of the Ikegami, the pedestal adjustment of the Sony was lowered from 25 to 15. The detail was also raised from 12 to 35. Although the horizontal and subcarrier phase had been adjusted, the detail and pedestal should be adjusted so differing brands may be identical.

Now both cameras could be cut together with very little difference. Although not switched live, the cameras would record their separate images on tape. In editing, the best angle would be selected.

#89 How to React to Something That Isn't There

Trained actors are great at reacting to something that is next to them, but what happens if the other talent or effect is digital? This requires a little bit of forethought and planning.

I worked on a bedding video where the talent needed to interact with onscreen graphics. This is a true test of an actor's ability. I have a hard time reacting to something that's in front of me; our talent had to be able to interplay with something that would be created in postproduction.

In one of the showroom scenes while speaking to the camera, the actor sat down on one of the beds, and graphic prices added up next to him. Eventually, these prices got so high that they knocked him off the bed. The character-generated prices were created by Adobe After Effects and Speculator's Infinity.

The actor felt like a nitwit talking to the camera and then falling off the bed. We actually had to do this several times to get the timing of the fall correct. Only professionals will fall off the bed time after time.

The other thing to remember is that your talent must be looking in the right place or the effect is lost. I worked on a Japanese HDTV shoot when they used a cardboard stand-in for all the digital shots. This involves a little more work in post removing the cardboard figure, but it does give the actors a realistic reference point.

This also helps with scale. If you tell your actors they are reacting to a three-foot gnome, do they really know how tall three feet is? If an object of the approximate height is placed in their line of sight, they will have an easier time of it.

#90 How to Use Subjective Shots Effectively

#90

Nothing is more exciting than a point-of-view shot of something that is quickly moving close to the ground. Subjective shots enable the viewer to identify with the object in question, so they should be a part of every shoot. I've done more than my share of subjective shots, but how do you realistically portray something like that? The easiest way is to actually place the camera in that object and then make it move. Voila! Subjective shot!

I needed a subjective shot of a wagon skipping its way down a wooded path on its way into a stagnant lake. I was chosen (short straw again) to sit in the wagon with a $160,000 camera on my lap. If you would like to recreate this feat, remove all of the nonessential camera equipment (extended viewfinder, follow focus, video assist, bumper stickers, and so on) that isn't needed for a shot of this type. Even a BNC cable attached to a faraway monitor will do nothing but get tangled at the wrong moment. Use the widest-angle lens you have access to and keep the camera as low as possible.

Your body will absorb some of the shock of the movement, but the viewer needs to see this jostling to feel the effect. This is why subjective shots of roller coasters make you feel as if you are there (the slight movement and camera shake). If your video camera has image stabilization, turn it off for a shot like this. You *need* the camera shake.

With my feet comfortably folded behind my back, I was able to get the front lip of the wagon in the frame. This gave the viewer a point of reference so they can say, "Oh, yeah, this is what the wagon is seeing!" This isn't mandatory, but it's nice if you can do it.

As I was pushed (in the wagon), I started the camera. Four seconds after the start of the shot, it was all over. My massive 165-pound frame and the 35-pound girth of the camera proved too much for the little red wagon. Three seconds into the shot, the right front wheel snapped off and the wagon lunged sharply to the right as the axle buried itself in the dirt. I never had to admit this before, but this is the first time I ever fell off the wagon! Figure 9-5 shows the camera and the little red wagon before the incident.

As a competent cameraman, I kept rolling (also because I had no idea what had occurred; "Officer, it all happened so fast!") What I mean to say is the camera kept filming as I went rolling.

As my knees were implanted in the soil, I never remember the camera hitting the ground. My camera assistant, Greg Ressetar, grabbed the running camera from my hands as I became one with the ground.

Always have a contingency for the camera. When you are placing any camera in potential harm, make sure someone is there to grab it if it should fall. They always said I could be replaced; the camera could not.

Figure 9-5
Wagon-cam with a very expensive wagon, I mean camera

In a second attempt at this subjective shot, Greg and I each grabbed one of the camera's handles and ran through the woods like two Groucho Marx caricatures holding the camera three inches above the ground (the same perspective as the wagon). The handheld jostling of the shot actually looked more realistic (safer) than me sitting in the wagon of death.

The key to a shot like this is not to dwell on it too long in editing. The viewer just needs a brief glimpse of the action to understand what is occurring (a second or two). If the shot is on too long, you've given away the effect. If using this run and gun approach, remember to keep the camera at the same level as if it were the subjective item. Any twist and turn Greg and I made while running would simulate the wagon hitting a bump in the road.

People expect to see a bump occur on the screen when they see the object causing the bump. In *The Shining*, the little boy is riding his Big Wheel through the hallway of the hotel as we view it subjectively from the his point of view. Shot with a Steadicam (a free floating gimble unit that eliminates bumps), the viewer notices a threshold approaching. When the tricycle flies over the bump, the Steadicam smoothes it out: no camera bump. The theater audience will laugh as this occurs, expecting a slight camera bobble to transpire. If you want the viewer to experience what the camera is seeing, if a bump appears, let them feel it.

#91 How to Show TV on Video When You're Not

We've all seen the effect in a movie where someone is watching TV and we can see the images of lighting dancing on their face. The same thing can happen if you are shooting in a movie theater and want the projected image to appear to flicker on the blank faces and stares.

The realists out there can actually have an image playing on a TV set, but the viewing audience will never know that is happening unless you show them the TV itself. But if you are impressed with the illusion of that concept, it's rather easy to create the effect without ever showing the TV.

Light your set as you normally would for either day or night. The simulated light from the television will be emanating from its own light. Set up a 600- or 750-watt open-faced light and gel it with amber gel. Place this light in front of the talent to simulate the light from the TV screen. The victim needs to look in the general direction of this light but not directly into it; you don't want him or her going blind.

You must also remember two things. The first is to set the height of the light at the intended location of the TV screen. If the TV is set on a dresser six feet off the ground, the light should be six feet high. The second is to simulate the size of the screen by using the light's barn doors. A 13-inch set will only throw light a small distance, whereas a large projection screen will cover a lot more area. The time of day also makes a difference. At night, the cathode image will travel much farther and fill the room with ambient light, even from a small TV set. In daylight, this effect will be much more difficult to see, so either the TV light must be brighter (a 1K) or the ambient light of the room must be darker.

Have one of your able-bodied assistants slowly move his or her fingers in front of the light to have shadows dance across the talent's face. Faster movement of fingers means an action scene and slower movement should be for a love scene. Practice this movement until you have the desired look. Have your assistant continue this effect until you start to smell flesh burning.

A simulated movie screen is not any more difficult than a television set. Movie theatres are always dark places because the image is projected, but you don't want the flickering light to be too bright. TVs actually cast a lot of light onto the viewer because of the cathode picture tube. The image of a movie screen would never be that bright, even in the first row.

Have your talent look up towards the light and diffuse it heavily. A soft-light works best because it most resembles a projected image. Fingers won't burn as much moving in front of this light, but anything waved in front will do the trick.

Chapter 10

Specialized Shooting Challenges: *Shooting Moving Objects*

Solutions in this Chapter:

#92 How to Shoot from Moving Objects

Shooting has its own set of inherent difficulties, but when you must shoot from a moving vehicle, that adds its own set of problems. This chapter will discuss the best way of shooting from a variety of moving objects.

Starting off with things high in the stratosphere, to getting your feet wet, to finally planting yourself on solid ground, I'll relate how to shoot from a hot air balloon, airplane, helicopter, boat, motorboat, sailboat, train, car, and through a windshield. The chapter also discusses how to use car mounts.

#93 How to Shoot from a Hot Air Balloon

Shooting footage from a hot air balloon is not difficult, but it does take a unique approach. Unlike shooting from other moving vehicles, you could burn out in reentry if you fall out.

The best time to shoot is usually early in the morning, soon after the sun rises. If you shoot in the afternoon, the haze in the air (from pollutants) is much heavier and more difficult to shoot through. Evening shoots allow you to see extremely long shadows of everything, including the balloon and yourself. Therefore, morning has less haze and the sun isn't in the right position to cast too many long shadows.

The only drawback to this time of day is the presence of fog if you are shooting in the spring, summer, or fall (winter has its own set of problems, besides you icing up). Because the ground is cooler than the surface air, a misty fog usually clings to the ground until about 9:00 A.M. Once this diffusion burns off, you will have clear sailing.

Balloons are not steerable, so if you want a steady hovering shot, a backward angle, or a Dutch Tilt, you are out of luck. The wind determines where the balloon will go at any time and the pilot will guide it. Of course, you are moving much slower than an aircraft, allowing you much more time to capture the images you desire. Any kind of camera movement will need to come from you. Image stabilization isn't really necessary because very little will cause any shaking.

You probably will need to travel higher into the stratosphere than you anticipate in order to shoot. I thought 400 to 500 feet would be high enough to capture sweeping vistas. You will get a vista, but if you want sweeping, you need to be up around 1,400 feet.

At any time of year the temperature is cooler at higher altitudes, so dress accordingly. Your batteries also will not be as cooperative in this colder climate. Any movement in the suspended basket will cause it to shake. You won't fall out unless you deliberately try to jump, but to some, the swaying of the basket feels weird.

When your balloon lands, expect a jolt when you suddenly hit the ground. Because the ride was so tranquil, I never expected the landing to be so harsh. Just bend your legs when you come in contact with terra firma.

Your flight will be extremely smooth without the jarring travel of a helicopter or the turbulence of an airplane. If you desire a steady, slow, fluid movement to your travels, a hot air balloon is the only way to fly.

#94 How to Shoot from an Airplane

Somewhat like shooting from a helicopter, videotaping from an airplane is different enough that you should be warned ahead of time what to do and also what not to. If you need steady, upside down, or backwards footage, or you need to land on someone's roof, the helicopter is your best choice. But if you need sweeping aerial views and have less money to spend, an airplane could be your ticket (an air joke).

The same concepts that I mention in the helicopter section apply whenever you are shooting in airplane, but several differences exist as well. Airplanes need to constantly move or they will fall from the sky. If the engine stops, you can glide to the ground (sometimes), but usually your shots are smoother (less shaky) than those in a helicopter.

Once in the air, focus on the job at hand. The windows in an airplane are much smaller than a helicopter's and are usually closer toward the ceiling. When you find the best vantage point, get as close to the window (or opening) as possible. I was so nervous in getting the shot that I kept one eye glued to the viewfinder and the other eye closed tightly. It was quite a feeling as I watched the tiny objects getting smaller as we rose into the great blue yonder. After a while, I didn't feel so good. All the blood had left my head and any food in my stomach was about to quickly vacate.

I was sure that the last thing I would hear would be an angel singing. Colors were forming and blending on the horizon. The icy blast of the air froze the eyepiece to my cornea. (They say you get warm just before you freeze to death; I was now quite warm.)

As we approached Mach 1, I started to black out. My hair had a permanent three-inch-wide part, and the gray hair count had increased twelvefold. Fearing that I was going to burnout in reentry, I held the camera as my knuckles began to bleed from the G-forces.

After I glued my beard back on, I realized I had just shot the most exciting visual image of my life. I also learned that bugs actually scream right before they splatter on your lens, but for some reason I still felt very ill.

After I asked the pilot where the nearest bathroom was located, he noticed I looked and felt like a department store mannequin. Sliding out of his seat he said, "Here, you fly the plane." I had to beg permission from my mother to let me drive her 1964 Chevy station wagon and now I was asked to fly a Piper Cub. Where's the justice?

Grabbing the wheel with clammy claws, my vise-like grip wouldn't allow the wheel to wander more than a millionth of a degree in any perceivable direction. When he switched off autopilot, I was muttering a final prayer because we both were about to die.

"Have fun with it," he suggested. Did he want me to have fun with both our demises? He gestured to climb and bank to the right. Over the course of the next 22 minutes, I managed to pull off that maneuver. As I was panting wildly from my hair-raising brush with aerial death in my loop-the-loop, he explained why he asked me to fly the plane. I knew he was just getting his story straight for the massive inquiry with the press.

By giving me the wheel and a chance to fly solo, he was taking my mind off my airsickness. Because I was so worried about keeping the plane up in the air, I had forgotten about my nausea. He knew I couldn't hit anything at our altitude, so steering would give me a feeling of control over the plane (Ha!), and someone who is flying an aircraft for the first time rarely pushes into a nose dive or sends the occupants directly into the sun.

This actually does work, but if I had kept both eyes open I wouldn't have been sick, and I also never would have flown a plane. I won't tell you how difficult it was for him to get back into the driver's seat; I was now headed to the Caribbean.

#95 How to Shoot in a Helicopter

Once again, you can shoot from a helicopter in one of two ways: the inexpensive or the normal way. Since most people fall in the previous category, that's what I'll discuss. Besides, how thrilling is it anyway if you have a Tyler mount on the front of the helicopter and are controlling the shots with a remote control? Isn't it more exciting to be risking life and limb by hanging out on your own?

Before getting into the helicopter, have a good idea what you are required to shoot: wide, sweeping vistas, complicated aerial maneuvers, extreme close-ups of stuff far away (next to impossible), or the sweat pouring out of your own body.

My instructions were to shoot (from the air) miles and miles of trees and open farmlands. The shot's perspective would be just a few feet above the tops of the trees. As the camera flew over this breathtaking vista, a mountaintop would loom in the distance. The brave soul in the aerial device would then slowly lift over the top of the trees to reveal the valley below.

Find the lightest, best-quality camera you can get and use image stabilization if it boasts that feature. Everything, including your knees, will be shaking, and this will eliminate some of that movement. A smaller camera is far more comfortable to hang onto during the flight.

Shoot with the widest lens you can find or get a wide-angle adapter. Besides allowing more information into your frame, the wider the shot, the less shake you will observe.

Another problem you must solve before being aloft is how to hold the camera. I asked myself (in my mind, not aloud as if I were talking to myself), "Should I hold the camera on my lap and frame it through the viewfinder or hold it on my shoulders and let my body absorb the movement?"

Discuss with the pilot what types of shots you'd like to get. Let him or her worry about flying the helicopter; you just keep your shot framed and in focus. I don't believe the pilot wants to die any sooner than you do, so don't dwell on how he's going to pull this shoot off. If it's something that can't be done, you will be told.

Carry extra everything with you in the cockpit. It's extremely expensive for someone to throw up (a bad choice of words) an extra battery as you fly by. Having spare tapes, batteries, and life insurance will save unnecessary stops. Cockpits are small to say the least, and try to keep your lateral movement to a minimum. If you accidentally bang into some knob or lever, you better have a great alibi if and when you land.

With all the noise once in the air, carrying on a conversation with the pilot is impossible without shouting. Unless you wear headsets, develop signals to speed up, slow down, climb, descend, and "stop, I want to get off."

Make sure you are strapped in, and record every second of your flight. Tape stock is cheap compared to the expense of the helicopter rental and you may find use for some of the takeoff and landing footage. En route, while shooting, determine the best vantage point for the camera. Buckle yourself in tightly and be prepared for gale force winds if the door is open or closed.

I was able to get great shots with the camera on my lap, but my knees knocking added too much movement. As I hunched over the viewfinder like Quasimodo looking for a lost contact lens, droplets of sweat rolled off my nose onto the black and white image. My new "for distance only" seeing glasses were quickly becoming a pair of granny-style reading glasses as they slid down my perspiration-lubricated proboscis. I determined that shoulder mounted was definitely the way to go. The human body does absorb most of the shake, even mine. Like Don Knotts at a social function, I was shaking like a Chihuahua on a December morning.

Once the camera was on my shoulder, the shot actually stabilized. I kept my nonviewfinder eye opened to judge the horizon so that I wouldn't drench the pilot with the remains of my burrito lunch.

If you intend on shooting people from the air and want to get natural reactions, make sure someone on the ground knows what you are doing. Upon hearing a helicopter, all people on the ground will stop, look, and stare. On a good day, some will even wave madly at you like they've known you for years (or they are signalling that something on your helicopter is about to fall off).

It's very difficult to hold a camera steady when a helicopter is hovering. The slower the air speed (we were stopped), the more unsteadiness in the toy. The wind is now pushing you around as the pilot is trying to keep the aircraft as motionless a possible. Don't try this at home.

Try to enjoy the experience. I doubt you'll get airsick because you will be too focused on what you're doing and on, as the Bee Gees said, "Stayin' Alive."

#96 How to Shoot in a Boat

I always seem to get the jobs of shooting from some kind of moving vehicle. You've read my exploits in the air and on the ground, but what about in the water? Shooting footage from a boat is a skill that someday I will master. Not wanting to delve into jet skis, Ski-Dos, or speedboats, I will be discussing motorboats and sailboats.

#97 How to Shoot in a Motorboat

As the name implies, these watercraft travel a bit faster than their wind-blown cousins. Covering larger distances at a blink, the motorboat will get you from point A to point B in no time. Shooting from an ocean liner or cruise ship isn't much different than shooting from land, but smaller and bumpier motor boats can loosen your teeth.

The faster your ship travels, the rocker the ride. It seems like the surface of the water is full of potholes because as you reach the crest of one wave, you will crash down over the next. Trying to hold a camera up to your eye is next to impossible at higher speeds without losing either your nose or the camera's eyepiece.

Image stabilization is critical in a fast-moving vessel and the camera should be held as close to your body as possible to absorb most of the shake and impact it will receive. Be more concerned with wider shots on the water because telephoto lenses and close-ups will be unusable. A camera stabilizing vest like a Steadicam or Glidecam will soften any of these visual obstacles.

If you can anticipate when the boat will hit the water again, you can loosen up (rather than the natural stiffening up for impact) and your shot will be more fluid. If you are shooting another water taxi, it's usually best to do that from shore if you need telephoto capabilities. I was videotaping an ex-president returning from a fishing trip in Maine (no names, but it rhymes with Bush) and was shooting at my most telephoto position. If shooting from another boat, I would have seen nothing but blur, but due to the steadiness of land and my anticipation of the direction of travel, I got the shot (without getting shot).

#98 How to Shoot in a Sailboat

There is nothing more romantic, serene, quiet, and nauseating than the majestic sailboat. These ships will crash up and down like a motor boat, but because of their girth, the effect is normally much smoother. I've shot dozens of water sunsets and the sailboat is the only way to travel.

Just like shooting in a motorboat, keep your camera wide and keep zooming to a minimum, even though you may have smoother sailing. With any moving object you may be shooting from, try to frame a piece of the boat, aircraft, or car in the shot unless you are trying to hide that fact. I always enjoy getting B-roll of the rigging, masts, and sails to intercut with my other footage.

As with any moving object on land, sea, or in the air, keep both eyes open and don't focus only on what you see in the viewfinder. I promise, with

Figure 10-1

A view from a sailboat (Photo courtesy of Linda Gloman)

the slow rocking, the horizon moving up and down, and your newness to this event, you will get sick.

If both eyes are open, your brain will realize that you are in a boat and keep your last meal from surfacing. Enjoy the experience with two eyes and you can replay the video from your stationary bed. Figure 10-1 shows the relative calmness and steep angle of shooting in a sailboat.

#99 How to Shoot on a Train

Since *The Great Train Robbery* (1903), people have been shooting on trains and loving it. It was now my turn to experience some of the fun.

It's amazing the clout a film production company has. With a wave of the assistant director's hand and a few words on his walkie-talkie, the engineer of our train would take the crew anywhere it wanted. Our train was a steam locomotive, but the principles here will apply to any train.

I checked out the interior of the locomotive during a break. I saw no automatic transmission gearshift that the engineer could send into reverse to try another take. Each time the train moved forward or backward, the engineer would have to do quite a lot. The steam engine was "idling" all day; after 20 minutes of no movement, the built-up steam pressure was automatically released through a vent. Talk about a loud sound. I'm still pulling briers from my backside when I landed in the bushes on its first blast.

Our entire day would involve shooting interiors in one of the train's passenger cars. At the front was the 1909 steam locomotive. Directly behind was the coal car to feed the hungry behemoth. Behind the coal car was a train car with a diesel generator to power our lights and video assist. Car number four was a passenger car that carried 93 extras. The last car, number five, carried all the lights, equipment, crew, and stars.

Since the passenger car had rows of windows on both sides, Kino-Flo units were chosen to supply the needed daylight-balanced fill. Attached with C-stand arms and poles, the Kino-Flos were silked and hung over the antique brass fixtures on the train's ceiling. In order to preserve the wood's finish, tennis balls were attached to the back ends of the arms. The noiseless, flicker-free, and cool Kino-Flos matched the daylight to perfection. The solid mahogany walls, ceiling, and trim were never seen because the action involved the actor looking out the window.

Over a six-mile stretch of track, the train would travel from end to end all day and we recorded scene after scene. We had AC power for the lights and video assist, and we traveled in self-contained comfort. The soft, green mohair seats were platforms for the Steadicam, camera, tripod, and crew, as well as video assist. The real beauty of the train was missed as our car housed us and the two cameras only saw a small portion filled with extras. Some behind-the-scenes shots are shown in Figures 10-2 and 10-3.

Because the director had hired a Steadicam in addition to the first unit camera, I now had two monitors recording different images. The director

Figure 10-2

Life aboard a train

Figure 10-3

Equipment and railroad car

monitored the Steadicam as the director of photography watched his crew on the other monitor. It was relaxing traveling in turn-of-the-century comfort with my video technology comfortably powered by AC. I didn't matter to me how many times the train traveled the same section of track. I was still amazed that someone had the authority and power to make an antique mode of transportation work its steam tail off. I spent the rest of my free time picking unburned coal particles out of my hair.

When we finished shooting the 12 hours of interior footage, I was responsible for packing up my video gear. Within two seconds of "that's a wrap," the lighting people dismantled all of their equipment. Like locusts, the train was stripped of movie equipment in 30 minutes.

I nearly had almost all of my video equipment back on the cart; I still had to retrieve the 13-inch monitor from the train seat. I climbed back aboard the empty passenger car and walked to where the monitor was resting. As I carried it to the door, I heard a very unpleasant sound from the lips of the engineer. "All clear!" I thought all clear meant no one is left on the train. I was still inside with my monitor and it wasn't "all clear."

I glanced up at the open door at the end of the passenger car. In my absence, they had removed every other car except the one I was standing dumbfounded in. Not more than 15 yards in front of me was the rear end of the coal car being pushed by the steam engine. I only had milliseconds to think before I would be a permanent spot on the back of the wall. I planted my foot as firmly on the floor as possible and gripped the monitor close to my chest; at least we would die together. I envisioned reading that on my tombstone, "he died with a video monitor clutched to his bosom." With the force of 12 tornadoes, the coal car slammed into my occupied passenger car and sent me tumbling back five rows. I staggered to my feet looking for shards of glass embedded in my torso. Fortunately, we were both still intact. I have never been hit that hard in my life and that linebacker in high school was no comparison.

Life aboard a train is cramped to say the least, but as long as you have sufficient power for your equipment, know where everything is, and what you're supposed to do, it's a breeze. It's just like walking down the tracks.

#100 How to Shoot Through Windshields

Windshield shots present several problems that need to be addressed individually. When shooting through any curved piece of glass, reflections will drive you mad. You will see yourself, the camera, the sky, and everything but what is happening inside the car. A Tiffen polarizing filter should be attached to the camera and rotated until most of the reflections are eliminated (there's never a good Terminator around when you want one). However, the camera's image might still be visible in the shot.

The camera may be covered with a black Hefty trashbag, but strong winds may cause the bag to ruffle, flip, and fly into the shot. A blanket can be employed, but its color will cause a reflection in the glass, so choose a dark, unobtrusive color. If all else fails, use the gray trunk liner and tape it over the camera.

This works well when the sky is overcast, but the split second the sun peaks through you will be scrambling for more blankets. Tape these additional blankets over the camera(s) and windshield with extended C-stand arms or sticks from a dead tree. Figure 10-4 shows how dead branches can have a second life. The edges of the blanket should be taped to the roof and tucked into the windows that can't be seen by the camera.

Depending on the sun and the exact angle of the camera, you may get the shot you're after. Experimentation is the only way to determine what works best.

#101 How to Shoot Cars

I always jump at the opportunity to shoot a car commercial. I have done a number of them over the years and no two have ever been alike. Car commercials are not very difficult, but I still like to strap on my thinking cap and come up with a new twist.

Figure 10-4

Can you find the car mounts in this shot?

A local Chevrolet dealer called me and said he wanted a spot featuring some of the less popular used cars (or should I say "preowned") that he had on sale that particular month. This was nothing out of the ordinary. He also wanted to have a spokesperson selling the cars on camera. Once again, that didn't seem to be an unusual request. Everything he told me over the phone was typical of what I had done for them in the past as well as for other dealers. The shoot was scheduled and I looked forward to the big day.

In anticipation, I got there two hours ahead of the scheduled start time to set up. Since the showroom was filled with new cars (mostly convertibles), every new model had to be removed and the preowned units brought in. With a set of magic keys, one of the salesmen opened the glass wall that separated the showroom from the parking lot. Suddenly, as if on cue, the sky opened up and a torrential rainfall began to saturate the ground.

In order to keep the new convertibles looking that way, their tops had to be raised. I've put up many a convertible top in my day, but this Camaro had to be one of the most difficult. In order to help the salesman from losing one of his appendages, I offered my assistance. For the next 15 minutes, we heaved, pulled, stretched, kicked, and yanked on that top. It wouldn't attach to the top of the windshield. I suggested that we also offer this new car as a preowned car for the sake of the commercial, but that fell on deaf ears. Finally, after every employee and customer in the building tried his or her hand at it, the top was attached. The Camaro was swiftly sent out into the rain.

The rest of the vehicles in the showroom posed no problem. Each pre-owned vehicle was then brought inside. Once inside, they were dried off, a piece of plastic placed under each tire to insure the floor wouldn't be marked, and the cars were slid into position. When I say slid, I mean just that. It's amazing what you can do with a 4,000-pound object and a bit of muscle. None of the vehicles could be started once they were inside the showroom because the carbon monoxide fumes would linger. Each car would be started out in the rain, driven at breakneck speed until it reached the lip of the showroom partition, then the engine would be shut off, and the car would coast to its spot.

This also worked quite well until it was the four-wheel-drive Blazer's turn. As luck would have it, the behemoth got stuck on the lip. An overanxious salesman threw it into four-wheel-drive and put the pedal to the metal. With an ear-piercing screech, the vehicle hopped up over the lip and went sailing across the room, narrowly missing its less expensive siblings.

In order to fine-tune the lighting, each car had to be positioned just right. Originally, they had the arctic white Blazer next to the jet black Camaro, which was next to the fire-engine red Corvette. Watching these cars being rearranged was like attending a behind-the-scenes rehearsal of a Busby Berkeley musical. With a snap of his fingers, the general manager and 20 salesmen orchestrated *Swan Lake* in the showroom. With a twist of his wrist, a car was moved two inches to the right, left, or across the room. Bear in mind that this was all accomplished without mirrors, starting a single engine, or moving a piece of plastic under their wheels (the cars', not the salesmens'). To this day, I still don't know how they managed that task with such finesse.

#102 How to Use Car Mounts

Car mounts need constant attention, they have to be adjusted constantly, and the second you turn your back on them, they fall off the car.

When using car mounts, the car should be towed. Although the driver's visibility may be obscured with the camera mounted on the side, no policeman in his right mind will allow you to drive with a car mount on the front.

The side and back windows should be gelled with ND9 gel and be taped to the car's trim. This will cut the daylight filtering through the already tinted glass and soften the harsh light by two stops. Lights should be used inside the car to illuminate the actors, which will be discussed in another section.

Camera mounts come in two styles: hostess trays and hood mounts. The hostess trays, like the name suggests, are mounted on the side of the car to capture a view from the driver or passenger's point of view. These mount anywhere on the side of the car and can be clamped to the door or wheel well, like the tray hostesses use at drive-ins.

The hood mounts are meant to showcase the talent in the car's interior. You can always turn them around and shoot toward the road, but the hostess trays clamped on the car's side show more of the car's fender or wheel, making it a more pleasing shot.

These are the only types of car mounts that I've used. Without strapping someone to the side of the car, these mounts effectively hold any size camera where it needs to be. The hostess tray can easily be mounted by one proficient grip, while the hood mount, because of its size, needs someone on each side of the car to gently lower it into place. Also be sure to develop a code system of knocks to alert the driver of the car trailer what you need them to do.

With a car in tow, you are unable to correct flaws immediately. If you see something wrong, it will involve much more than simply reaching over and adjusting what's causing the problem. The towing vehicle must stop and you must then jump out without tripping over the hitch, stop the camera, and then address the particular problem.

Solutions in this Chapter:

Chapter 11

Specialized Shooting Challenges: *Shooting Locations*

#103 How to Location Shoot

You sometimes may find yourself out of your element when on location. This chapter will discuss dealing with some unique shooting locations that take special planning or, at least, everything working for you rather than against you.

Since you need to be in charge of your location rather than the other way around, how do you shoot in a glass room, in a health club, in a room surrounded by mirrors, in the driving rain, on scaffolding, at night, or in the tropics? Each of these questions will be answered, and I lived to tell about it.

#104 How to Shoot in a Glass Room

I had never seen a showroom built entirely of glass before, but I had the luxury of shooting a furniture spot in this room where all four walls were smoked glass. From floor to ceiling, all you could see was eight feet of glass. If one of the little pigs had built his house out of glass instead of straw, the wolf wouldn't have blown it down. Not only was every vertical surface of the room transparent, but the ceiling was flat black. I think lighting in outer space would have been easier.

The smoke coating on the glass showroom cut the light level outside by two stops, but we would still be shooting in the afternoon when the sun would be bearing down on the windows. I took a spot reading out at the faded asphalt parking lot that surrounded this glass showroom and got an F11 reading. The interior of the showroom was a whopping F1.8.

On the day of the shoot, we arrived with four rented 1,200 HMI lights. The focusing HMIs were spotted directly toward a sofa from a distance of 20 feet. The Par HMIs were silked with a four-by-four silk held in a fiberglass frame that softened their punch.

The focusing capabilities of the HMIs proved invaluable. The darker-colored leathers would absorb so much light that we didn't have to diffuse the HMIs. When we wanted to light individual pieces of furniture and make them stand out against the blinding sun coming through the glass, the HMIs worked perfectly. By slightly closing a barn door, we were able to illuminate the black leather sofa while keeping the beige sofa next to it from glowing (no relation).

In order to hide the chrome seams on the glass walls, plants were strategically placed behind the furniture. This gave the glass a seamless look. We didn't want to hide the fact that we were shooting in a showroom, but the reflections on the chrome strips that separated each glass frame were distracting. By angling the focusing HMIs, we used the shadows cast by the plants to add textures of light to the smooth leather surfaces.

On the other hand, the Par lights were excellent key lights. With their initial punch of light, I was able to bring the light level of the room up to F5.6. Since the outside illumination was still two stops over, I gelled the windows with ND6 gel. Three 28-foot rolls later, we were ready to shoot. As you looked at the sofas in the showroom, the light outside was the same level and color temperature, all in all a very even match.

As the people came onto the set, we blocked out each of their movements. If any person moved one additional foot to the right or left of the marks, he or she would immediately have a summer suntan. Figure 11-1 shows the stacks of furniture and the wall of glass.

#105 How to Shoot in a Health Club

Shooting in a health club poses its own set of problems. If it weren't for the people, equipment, stifling heat, and mirrors, it would be like shooting anywhere else. In order to make our actress appear more dominant and John Wayne-like in the frame, each health club image was shot at a low angle.

To make the health club scenes appear more visually appealing, our lighting was more diffused, bounced, and less film noir looking. We also overexposed each shot a half stop to make our newly transformed actress less foreboding and more attractive. She also wore makeup to make her exposed flesh look less ghastly. After carefully weaving power cables between exercising patrons in order to get the best angles of our transformed talent, we were lit and ready to begin. Some of the patrons began to complain about the heat being generated by our lights; it was making

Figure 11-1

Glass walls as far as the eye can see

them sweat. Can you imagine that, sweating in a health club? I always thought that if you went to a health club it was because you wanted to sweat. Here we were providing a service to them, and they were complaining about the heat.

Our health club wasn't the most modern facility in the city, which was evident by the way most of the chrome parts of the machines were rusting. The tears in the fabric on the benches were cleverly concealed by the people sitting on them. Luckily, I didn't have to scrape any "Buy War Bonds" stickers from the bottoms of the machines. The more modern equipment was jet black, so it absorbed all available light. We would point a light at the exercisers and the black hole would immediately absorb all visible illumination.

All health clubs have rush hours just like most highways do. We were allowed to shoot in the club but were told to be finished before rush hour. At 3:30 P.M. precisely, thousands of overweight, underpaid, high-powered executives and high-powered wanna-bes came streaming through the door. Since we were on our last shot by that time, we were close to packing up.

I learned another thing about video lights: The split second you turn one on, a person's brain shuts off. I think a connection exists between video lights and dumb questions. As soon as the AC current is switched on and the lights illuminate, the dumb questions begin (I guess it has something to do with capacitors or resistors in the lighting equipment). You will be asked questions such as, "Are you videotaping?" or "What do the lights do?" One of my favorites is when someone is exercising directly in front of the camera with the camera's red light illuminated and they ask, "Am I in the shot?" I never really know how to answer a question like that, but I'm working on it.

It's also usually at this time that the owner of the club (the client) will saunter by and find out if his or her money is being well spent. Our client was the typical muscle-bound individual whose main vocabulary words were "Coach says," "Woman . . . pretty," and, if pushed to the limit, "Beer." He saw all the lights focused on a particular piece of equipment and the camera's image of the object in the color video monitor and asked, "Are you going to film that piece of equipment?" I didn't have the heart to tell him that we weren't shooting in film, but I had bigger obstacles to handle.

Ten minutes into the shoot he immediately became director and had a whole new concept of how he wanted everything shot. "Why don't you put a light over there?" he gasped as he pointed to the left.

I answered, "I thought you didn't want to show that piece of equipment."

"I don't, but won't that make the shot brighter?" he replied. I then tactfully explained the basis of Lighting 101 and that seemed to quiet him for the time being. I know that the video was his project, but we had gone over the storyboards in excruciating detail the day before. Everything would flow like clockwork. But once he appeared on set, he had forgotten everything that we discussed (maybe he looked at one of the lights).

One of the grips, whom I now consider my favorite, tried to distract him by asking about a particularly unusual piece of exercise equipment. This immediately made the client forget about me and begin to show the grip how the equipment should be correctly used. After he set the weights to their highest position, he began to exercise. At least this allowed us to continue on schedule.

#106 How to Shoot in Mirrored Rooms

Another valuable lesson I learned while shooting at a health club is to never block the mirrors that surround the four walls at most clubs. These mirrors are an important way of life for exercisers. Even though mirrors are extremely maddening for lighting people, the average health club patron wants to see him- or herself strain and sweat as he or she works out.

Our solution was to place our diffused light directly in front of one of the gaps between the mirrors. Each mirror was joined to another one with a half-inch-wide black gap. At a distance of 25 feet, this half-inch gap was wide enough to block a light stand. On the few occasions where our "crack" wouldn't work, we placed an exercising extra exactly where he or she would block the light. My belief is to use mirror cracks (every mirror has to have one some place) or use extras to block light where it should not be seen.

In college, professors rarely tell you how to handle a room with a thousand mirrors. My best advice is to light it any way you can without using directional lighting. Hard lighting tends to reflect harshly on mirrored surfaces. Once you have the lights exactly where you want them, use people and machines to your advantage. I was surprised at how a lighting instrument could be hidden by moving a 900-pound piece of equipment 2 inches. However, don't move the machines that have 16 tons of weights on them; I learned that the hard way (I don't mind being called "hernia boy").

I thought that if I were to illuminate our talent near one of the mirrors, I could also take advantage of some of the mirror's reflective powers (not healing powers). But any time you turn on a light, everyone's attention in the room goes immediately toward that light. It's like magpies collecting shiny objects. As children, you're told not to look into the sun. You would think that the same thing would apply to adults looking at video lights. Maybe they like the way their skin glistens in gelled, tungsten light. Whatever the reason, even without them knowing it, I was able to use these walking, exercising human reflectors as bounce and fill lighting for the video and that's a great reflection on me. Figure 11-2 shows how mirrors can be difficult to light.

Figure 11-2

A nice reflection on you with mirrors

#107 How to Shoot in the Rain

Maybe a more apt title would be "Singing in the Rain," but that has its own set of copyright issues. Besides getting wet, shooting in the rain isn't difficult. It's just that some factors must be overcome, before you are.

When I worked at a local TV station, one of our lead anchors was asked to carry the Olympic torch when it came to our area. This is obviously quite an honor, and when you are asked (you must be asked), no matter what the time frame, you usually gladly accept.

Our TV personality had an eight-minute stretch (not a warm-up, the time he was carrying the torch) where he would run through the streets of Baltimore (we lived and worked 80 miles away) proudly displaying the torch for the Summer Games. Since you have no control of when and where to receive this honor, we all sat waiting until we were told.

His moment of fame turned out to be 5:00 A.M. on a rainy morning. Because the sky was overcast, very little ambient light was present. The cold, driving rain was pounding down on our skin as our runner, "Brad," warmed up in his shorts. I was stationed in the back of a pick-up truck, covered in a tarp.

Rain is almost invisible in video (especially in the dark), but at this ungodly hour, everything else was invisible too. In order to see the runner (we could see the torch because it was lit), we positioned the lights low in the truck's bed, close to the rear bumper. Once they were on, the lights illuminated each droplet of rain and relatively nothing else. Low-angle lighting does not work in rain unless that's all you want to see. Since we did not want to highlight the fact that it was raining, we used a sun gun (a portable battery powered light that attaches to a camera) on top of the camera.

The sun gun lit the rain as well as our runner. Life was getting better, but the silvery drops of rain looked too prominent in the frame. This was Brad's one shot at fame and he didn't want to be upstaged by rain.

Unfortunately, the best way to light rain is not to light it at all. Low-angle, high-angle, or straight-on lighting will all catch the reflectiveness of the falling water and make it more visible to the eye. If you don't want to see the rain (as in our case), you don't want to light it from any angle. This is where ambient light is so important.

Still in the predawn hours, the orange-colored street lights gave us an F1.2, four stops lower than our lens could handle. By upping the gain of the camera to 18 decibals (a real grain fest), we had two additional stops of light, but the image looked nasty.

Knowing that Brad would be holding the torch like the Statue of Liberty, the light from the flame would give us another half-stop on his face and torso. At this time in the morning, a viewer wouldn't expect a perfectly lit visual. Two stops of underexposure (all the background) and a one-stop loss around his face would look natural.

When the event happened, I had little or no time to plan anything else. The background was soft and his face was only a half-stop under as he ran with the Olympic torch flickering. People expect the subject to go in and out of light when doing something like this. The rain would occasionally be seen when it caught some reflected light.

For the most part, although I felt the rain, it could not be seen because we had no top, front, or side lighting, just natural ambient illumination. Fortunately, rain would not put the torch out (that would have been another issue), and I actually won a broadcaster's award for risking life and limb in the rain. Only if Brad hadn't accidentally set my beard on fire . . .

#108 How to Shoot on Scaffolding

Sometimes the only way to climb the corporate video ladder is to make it yourself. Scaffolding allows you to scale new heights, but be prepared for a different view of the world once you're up there.

The scaffolding should be erected from the ground up (do you know a better way?). Four six-foot sections attach to a wheeled base that rests on plywood squares that serve to level the shaky beast. Plywood must be used as the platform for the camera and lights. Boards may also be placed behind you as back support.

The only way up to the top of the scaffolding is to climb it, and the higher the scaffolding is off the ground, the less stable it becomes. Additional lights and the camera will share the tiny platform on the scaffolding with you. Remember to sandbag everything down because the wind and any movement will send your equipment sailing.

This is the sturdiest camera platform that can be inexpensively erected or easily torn down. As far as temporary structures go, this is the best way to "get up there" and shoot. Although shaky, the platform will not tumble to the ground if assembled correctly. Position the structure as close to the subject as possible because any telephoto shooting will add to the shake.

Have protection (sun and rain covering) because you will be subjected to the elements much more at this altitude. Raise the equipment once you have scaled the heights and lower it before you leave. Figure 11-3 shows my precarious vantage point on scaffolding.

Figure 11-3

A lonely boy and his scaffolding

#109 How to Shoot at Night

Shooting video footage at night doesn't have to be a grainy experience. It also doesn't have to be a black and white (infrared) trip either. With planned lighting, any exterior location can be shot at night.

The effect of night can be faked if you shoot during the day, but it never looks real. Underexposing and shooting through a blue filter (80A) will give the effect of nighttime, but with the tools available, why not do it at night?

In an exterior location, don't light your talent like you would in the daytime, unless you want it to look like spotlights are pointed at him or her. Instead, light from overhead. The key (a lighting term) is to have the ambient illumination look natural. The video camera needs more light than the naked (or clothed) eye, but you want the shot to appear as if a street light or the moonlight is the only source of light. When lighting indoors, often the "practicals" (actual lights seen in the room that function) are revealed in the shot so people know where the "light" is coming from.

Lighting from above is the natural way to illuminate a night shot. The shadows will be heavier because the light is high, so you need to cheat slightly. Heavily diffuse (not confuse) a light and point it at your talent to fill in the shadows. The face light should be far enough away to evenly illuminate the face without throwing the telltale shadow behind them.

Some people like to add a touch of blue to the shot (an 80C filter), but I believe that calls too much attention to the filter. Since the ambient light around your talent is virtually nonexistent, the viewer will surmise that it's night. Or, use an HMI at night and balance for tungsten, the light now looks like blue moonlight.

Light every subject or item you want the viewer to see and let the rest fall off in darkness. If the talent is moving, do not move the light with him or her. Rarely does a street light or the moon light follow someone around. It is a good idea to move the talent's face light as he or she travels; this fills in the shadows to expose the person for the camera. The major light can come and go as he or she travels through it.

#110 How to Shoot in the Tropics

As you get closer to the equator, things start heating up, and that goes for your camera equipment too. Moisture is one of the biggest concerns when shooting in the tropics and it must be dealt with properly, or you may see what your "dew" light looks like when it's illuminated.

On a trip to the rainforest in St. Lucia in the Caribbean, I learned all I could beforehand on how to prepare my camcorder for the humid journey. Rain can start and stop at any moment in the tropics and the humidity is always close to 100 percent. If you think you know what mugginess is, wait until you get to a rainforest and ask your camera how it feels!

Once again, prevention is the best action you can take to ensure your digital camera will function properly. The first step is to cover and protect the camera from instant rain showers. You can purchase plastic camera coverings from a number of sources, but because of the tiny size of most digital cameras, a plastic bag works just as well. I used a large Ziploc™ bag and carefully sealed the top until the pretty color had changed to purple.

Depending on the size of your camera, a sandwich-sized or larger freezer-style bag will work. Although these plastic bags look optically clear, they are not. Don't just drop your camera in the bag, zip it up, and begin shooting. Instead, cut a hole large enough for your lens to stick out and keep the rest of the camera in the bag.

Since my camera was a larger three-chip unit, I used the freezer-style bag. I cut a hole in the bag out of which the lens could protrude. In order to keep the lens from sliding in and out of the bag, I attached my lens shade over the edge of the plastic.

Before you hermetically seal everything up, drop one or two bags of silica gel (desiccant) into the bag to help prevent condensation from forming. This is another great reason to cut an opening in the bag; the humidity will cause condensation without the silica gel. Your camera is now ready for almost everything (if you don't need to look into the viewfinder).

The first question that comes to your mind is, "How am I supposed to hold my camera in a slippery bag?" Yes, that is a slight drawback. You can easily see and operate any of the features (zoom, focus, exposure control, and so on), but you must do so through plastic. I had to make a few other modifications for my particular camera.

Cut another opening in the back of the bag to allow the viewfinder to stick out. I used a little gaffer tape to keep this from sliding around. Don't try to use the flip-out side monitor because the daylight will be too bright, it will shorten your battery life, and it *will* get wet. You can do one of two things with the microphone. I removed mine because I would be adding ambient sound later during editing and all I would get on location would be footsteps, rain hitting the protective gear, and the crinkling of me holding the bag.

If you leave your microphone inside the bag, you know what type of noises you will hear. If sound is critical, the microphone must also stick out of the bag or you should use a lavaliere mike and attach that to yourself safely and securely. You now have three of your most critical elements protruding from the protective covering. If it starts to rain suddenly, try to shield the camera the best you can, don't point it directly up, and, if you fall into a puddle, throw the camera to someone else.

The bag principle worked well for me and I was able to open the bag to change batteries or reach anything else the bag obscured. Customize the bag the best you can and it will work quite well. However, if you want the best protection, buy an underwater housing unit for your camera (insert the silica gel) and you're safe down to 1,000 feet.

Chapter 12

Specialized Shooting Challenges: *Shooting Objects*

#111 How to Shoot Difficult Objects

One nice thing about shooting things is that they cannot talk back if they don't like what you're doing. All they can do is sit there and take it. But certain objects are a little complicated to shoot and this chapter tells you how to deal with them effectively.

A computer screen will roll or have a scan bar if the camera is not in sync with it. How do you shoot something in a pleasant manner if it's a disgusting subject? How do you make money look "real" on digital video? How about shooting wine and meat, or how do you light and shoot something transparent? (If you want to shoot a ghost, ask someone else.)

#112 How to Shoot Computer Screens

I used to think that any camera could shoot a computer screen, but I was wrong. All computers scan images differently and it takes a camera up to the task to capture a screen properly. The Clear Scan capability is the only tool that will allow you to shoot a computer monitor's screen without having to deal with the rolling scan bar.

Apple Computer recently donated some laptop and desktop computers to a school district in my area. I was hired by the school district to create a video on what this new equipment did for the district. These computers would allow the 4,500 students from kindergarten through twelfth grade to become more computer literate. First-graders learned writing and typing skills, intermediate and middle school students shot and edited video movies, and the senior high students carried laptops to classes with them.

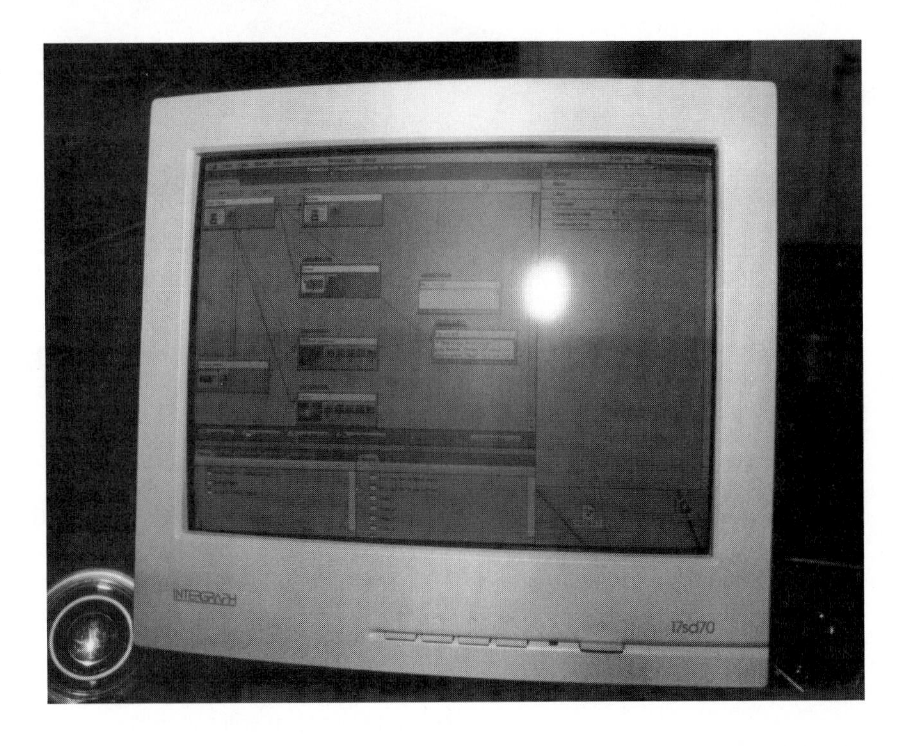

Figure 12-1

A computer screen with no roll bar

As eager students sat in the library, they worked on assignments using their laptops. Laptop computer displays, because they have *Liquid Crystal Display* (LCD) screens, can be shot with almost any video camera. The library was lit with two Arri open-faced 1Ks with opal diffusion and 1/4 blue. The 1/4 blue helped even out the flesh tones under the fluorescent lights. Two additional Arri 650's Fresnels also armed with 1/4 blue and opal were used as fill lights. Figure 12-1 shows a correctly photographed computer screen without the roll bar.

When shooting the desktop screens, we needed a different camera. The Mac desktops the students were using enabled the frequency scan rate on the monitor to be changed, but not enough to allow the camera to shoot a roll-free image. Since the client wanted their Apples to look good, a rolling monitor screen in the background would be distracting. As a solution, we rented a camera with Clear Scan, which enabled us to match the Mac screen rate. The Macs were adjusted to a rate of 95 hertz and our camera's Clear Scan was set the same way. This provided a crystal-clear, locked image on the monitors.

#113 How to Shoot Gross Things

We've all been in this situation before. You have a client who wants to make a video on . . . a sensitive topic.

One of the most difficult topics has to be medical subjects. Anytime you do a video for a doctor, you know you'll have to be careful about what you talk about and what you show. My motto always has been, "If you can't pronounce it, don't shoot it."

What kind of doctor do you think would be the most difficult to advertise on television? Yes, that was my first guess too, but the one I'm talking about is almost as difficult: a podiatrist. For those of you not up on your "ologists" or just plain "ists," a podiatrist is a foot doctor. I believe Dr. Scholls was the first. Most people don't really want to talk about their feet, much less show them to anyone. The podiatrist I worked for wanted to let the world (or local TV) know about how he fixes feet.

First, I had to come up with a concept that wouldn't have people running (a foot thing) from the room screaming. I always try to use a humorous approach (you never would have guessed that about me), but this doctor wanted no part of it. I told him the best way to handle a touchy subject like foot problems was to use humor. He said he didn't want to be known as a funny foot doctor. When you become an "ist," your humor slowly dissolves.

Since I struck out on my first attempt, a concept that I believed would work, the good doctor had his own idea. He said we should use testimonials (he didn't use that term, just like I don't usually use the term "hammertoe"). Testimonials could work, but it also presented a few problems (which he quickly made worse). He didn't want to use actors because they wouldn't be convincing. I tried to explain to him that an actor was someone who was a professional at being convincing, but he still said no. He wanted real people.

Real people *would* work because sometimes they add realism to a spot. He took it one step further saying he wanted to use actual patients. That's when I got scared.

When you go to the doctor, do you tell everyone what you're going to have done? If you have toenail fungus, hammertoe, or an ingrown toenail, do you tell anyone you have it? These may be painful conditions, but you keep them under your shoe (pun intended). I asked Dr. No if he thought any of his patients would want to talk about their problems on TV. He said they would, and that's where he proved me wrong. Everyone was so happy with the work he had done (remember I called him the "good" doctor earlier) they were thrilled to talk about their foot maladies on local television.

Since we had gotten past the first hurdle, we had to address the second. It was fine to have people talk about their problems, but we had to show something to the viewer. Luckily, all the doctor's patients had their problems corrected, so we couldn't show ratty, uncorrected feet. Because his budget was small (I've heard that before), we had to come up with a cheap visual. He wanted a visual that would show the problem without grossing anyone out (I won't tell you about the suppertime air play slot he chose). The only way was to create it graphically; a cartoon foot offends no one.

Graphics help take the realism out of the effect (if you want them to). If you used real bugs in *A Bug's Life*, it wouldn't have been as effective. People don't want to see real bugs; cartoon bugs (although very sophisticated) are more entertaining.

This was the beginning of problem three: how to make a cartoon foot not look like one. We would have to exaggerate the problem so it would stand out to the viewer. We would then show it corrected graphically (show a normal foot again), but he didn't want Bozo's feet to be used. I now had to do my homework.

I hired the most attractive models I could find and I shot their feet. The models would walk into our studio, stand on a gray Muslin backdrop, and roll up their pant legs. My boss called me aside and asked why I was just shooting the feet of such attractive models. I told him that this was the way my wife preferred I shoot it. After explaining to my colleagues that I hadn't developed a foot fetish, they understood my reasoning. I would get a normal foot on tape, and our graphic artist Stephen Brehm would create the "problem." It sounded simple, but it wasn't.

Gorgeous supermodels (okay, local women) look fantastic and have great bodies and personalities, but usually not much attention is paid to their feet. An attractive woman doesn't necessarily have good-looking feet; in fact, these didn't. I really enjoy shooting, but after looking at feet all day, I was ready to throw in the towel (I could have said shoe, but I spared you).

It took all of Stephen's artistic ability to make these models' feet look "normal." He grabbed a frame from our Betacam SP tape and used that as the background template. He then rotoscoped over the real foot and created a graphically good-looking foot. Once this normal foot was created, he had to create a "nonsex" foot. Before you start looking that up in your dictionaries, the doctor didn't want it to look like a male or female foot. I didn't realize there was much of a difference between the sexes in their feet, but I guess that's why I'm not a doctor (and a few other reasons). Once he desexed the foot, he added the "hammertoe." Complete with the painful swelling he created, I was able to dissolve between the hammertoe foot and the normal one. Figure 12-2 illustrates a tastefully done image of a hammertoe.

When all was said and done, the doctor really liked the spot. It presented the subject in a pleasant, nonhumorous way. By showing a problem many people have, the commercial would hopefully inspire people to get their problem corrected. After his client base grew to new heights because of this spot, I was ready to go back to shooting something I enjoyed doing, like

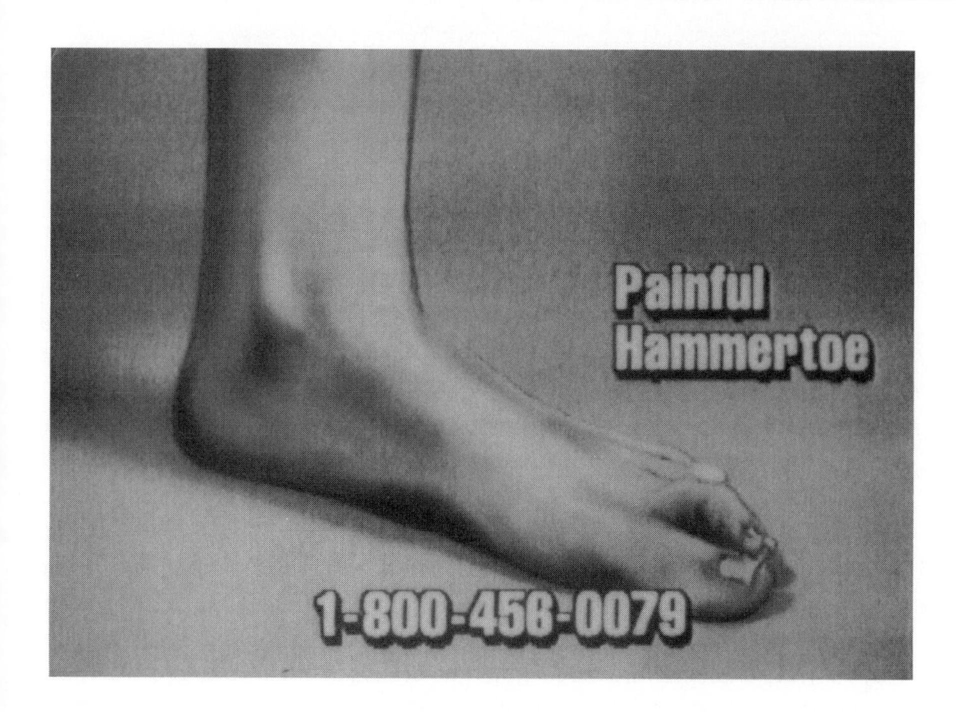

Figure 12-2

A hammertoe, politically correct

the parts of people we usually see. In fact, he was so thrilled with “Bride of Hammertoe,” he is now ready for his first sequel, “Attack of the Toenail Fungus.”

If feet weren’t enough, I was able to climb the corporate ladder and shoot knee surgery. In the procedure I would be shooting, two holes would be drilled into the knee, one to house the camera and cutting tool, and the other to allow saline solution to fill the area.

My camera in the surgery room filming the procedure would be intercut with the footage taken inside the knee with the orthoscopic camera. The best way to shoot something “real” (unlike cartoon feet) like knee surgery is not to dwell on any area too long. Shoot a variety of angles and images, avoiding extreme close-ups. Only doctors are interested in seeing this. The video I was shooting wasn’t for surgeons, so I never got too close to the action.

The toughest thing in the world to shoot, in my opinion, is eye surgery. I’ve done my best work passed out on the floor (mainly because I’m eye squeamish). This is something that needs to be recorded in extreme close-ups and I’m not the guy to do it. If you can handle it, think happy thoughts as you are shooting someone else’s eye.

Even childbirth is far easier to shoot (for me). Once again, don't dwell on close-ups of what is happening (unless you are instructed to do so). Instead, shoot cutaways of the husband holding his wife's hand or close-ups of the doctor, nurse, mother, and father. People know what's happening in the delivery room without you having to shoot another eighth-grade health hygiene video. By using your discretion, you can still shoot moving videos about topics some people would rather not see up close.

#114 How to Shoot Realistic Money

Have you ever wanted to show lots of money on camera but didn't have enough of the real stuff to make the shot work? The only option is to use fake money and somehow make it look real.

I shot a commercial for a bank who wanted an actor to be pulling a red wagon containing large glass jars filled with money, his life savings that he intended to bury. The glass jars of money were actually plastic jars stuffed with bubble wrap with fake money lining the interior. Real coins (because they're inexpensive) were used as small change, but the fake dollar bills were scanned from real ones, printed, and cut to the correct size. From a distance, these phony bills looked real in every way, and on camera no one could tell them apart.

On film, the color of the bills matched correctly, but obviously the paper was totally different (only the Treasury Department has the correct type of paper, so don't try to make this your new occupation at home). This cinematic counterfeit operation was brought into service because the money would be blown around, torn, soaked, and burned. Every time we would move the wagon, fake money would fall out of the jars. The money did look extremely realistic and all of it was destroyed after the shoot (sorry). In the wrong hands, our counterfeit money would lead to several years in the slammer.

To make matters worse, the red wagon would roll down a hill on its own, leaving trails of money, hit a bump, fly into the air, and end up in a lake. After each take, the escaped bills would be picked off the surface of the water, dried out, and used again in a subsequent take. We were finding lost bills weeks after the shoot. Just try your best to collect everything before you leave. Figure 12-3 shows how fake money can look real when you want it to.

We achieved the same type of effect with a "money tree" shot. The fake bills were attached to the tree, and on tape no one could identify the forgeries. If you will be working with extras or nonprofessional talent, make sure you use real money in all shots. Our fake money was used because it was destroyed after shooting. This is one calling card you don't want to leave behind. (I just videotaped the money in these spots. All of the phony money was made by others. That's my story and I'm sticking to it.)

Figure 12-3

Fake money in a jar (just like at home)

#115 How to Shoot Wine

Notice I said shooting, not snorting. Wine is one of the few things in video that can actually be shot with very little stage dressing. Right out of the bottle it looks cool, inviting, and intoxicating. But sometimes to have to add a little effect to fool the viewer.

If you want the wine to look chilled (it's a sin to serve red wine that way, so I'm told), add an ice cube—a plastic one. This ice cube never melts and it looks real. Just don't use the one with the fake fly inside. If chilled wine is poured into a glass, the heat of the lights quickly makes the glass frost up, making shooting difficult. Instead, shoot it warm and make it appear cold.

With a wooden skewer dipped in soap (or dishwashing liquid), a slight swirl will instill tiny, frothy bubbles that will cling to the edge of the glass (remember, these are soap bubbles and will last longer than carbonation). The only side effect of this skewer method is that a soap film will remain on the surface of the wine if it is not stirred occasionally (James Bond wouldn't like it). The same principle works with red wine, blush wine, or rosé.

Champagne on a limited budget is usually ginger ale, but with the same soap techniques to keep the bubbles alive longer. Illuminating the glass from behind or below causes the liquid to come alive with inner light. If you really feel daring, drop a small mirror into the liquid to reflect even more highlights.

I can't give away too many more tricks or the food stylists won't tell me any more of their secrets. (They would, but they'd have to kill me.)

#116 How to Shoot Meat

How do you make a lump of meat look enticing? The only way is to prepare it for the camera by using makeup: food makeup. Using hair dryers, hand steamers, Q-Tips, cotton balls, and surgical instruments, you can make your beef look beefier, your pork look leaner, and your crab . . . crabbier?

Let's begin with crabmeat. No matter how you slice it, it still looks awful in its naked state. When removed from the shell (really naked now), use an X-Acto knife to remove any fine hairs or fuzz from the meat. Pull or poke the various ridges and valleys in the lump to enhance the texture. Gravy coloring (the dye that gravy is mixed with) can be brushed on the top to create a crispy-like slight brown color. To add a bit of glistening, spread olive oil across its surface.

Being a light-colored meat, chicken needs a little extra help being bronzed at the beach. Do not shoot chicken raw because it does not look appealing. Instead, cook it slightly until all the pink is gone and the color is a light brown. Martha Stewart says the next step is to pull out some Miniwax Walnut Stain and lightly brush the surface of the meat (go ahead—use a foam brush). The golden-brown, freshly cooked appearance will last indefinitely (and so will the smell). Add olive oil for that extra sheen after the stain has dried. (I'm surprised she didn't recommend varnish or shellac.)

Steak looks great cooked on the outside while very rare on the inside. Somehow people find bloody meat more appealing (like rubbernecking at an accident). Quickly char the outside of the meat by painting it with butter, olive oil, or Italian dressing. Use a fire (blowtorch) and scorch the liquid until it begins to blacken the surface. Apply stain, gravy coloring, or soy sauce to the red (raw) portion and add a *slight* amount of water and red food coloring to simulate blood. You don't want pools of blood like a homicide, so use discretion.

Surgical instruments (dental scalers, scalpels, and forceps) also aid in fluffing up the meat before the big fight. Play around with a stand-in slab of meat before mutilating the star with a sharp instrument. Be careful when using these instruments or the blood you shoot may be your own.

#117 How to Shoot Transparent Items on the Floor

This may not be something you are called to do often, but if you ever need to shoot something transparent, a solution exists. How do you light (and even see) clear, dried glue on a hardwood floor?

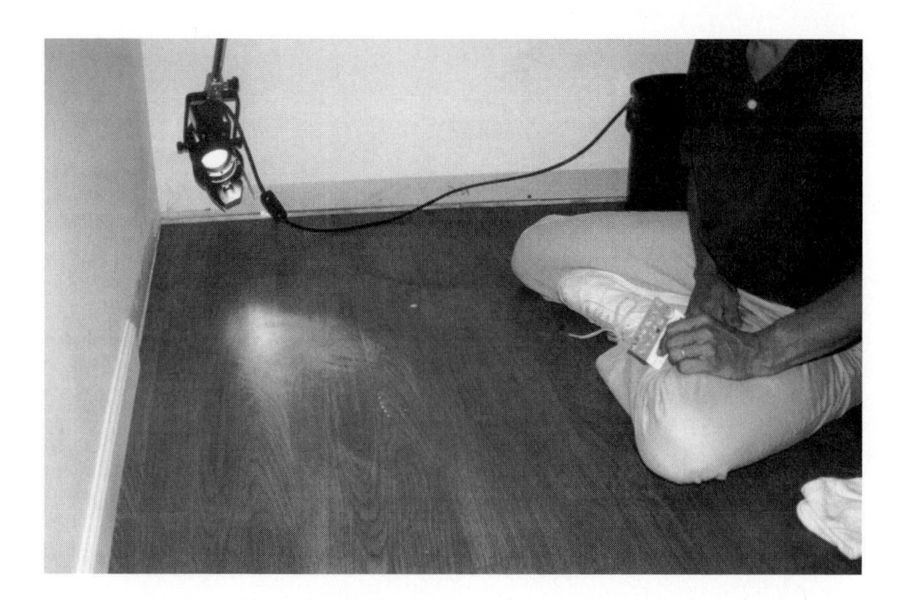

Figure 12-4

Dried glue, a light, and a camera. What could possibly be more fun?

On a particular shoot, once the flooring had been installed, we had to show how easy the cleanup was. Glue from any assembly that was left on the floor could be removed by simply pulling it off once it dried. This was easy to say but difficult to show. The glue applied white, but dried clear. We racked our brains to come up with a clever way of seeing the glue, but not overexposing the floor.

The only way we could figure out was to skim the surface of the wood flooring with light. A 200-watt Pepper light was attached to a C-stand arm and lowered to the floor. The 200-watt light was too strong, however, and had to be diffused. With a piece of 216 attached to the front, the Pepper became our glue-finding light. Figure 12-4 shows the best way to find dried glue on the floor.

Held at a 30-degree angle above the floor, the clear, dried glue could be easily seen by the camera. By simply scraping the edge of the glue, the end could be lifted and the dried glue removed. We were so excited by this technique that we created 10 test patches of glue and watched them dry (okay, it was a slow day).

The dried glue was supposed to be easily removed by sticking (or sliding) the edge of a credit card (an expired one) under the edge of the glue and lifting it up. As soon as the cameras started, the card trick did not work (pick any card; they all won't work). We actually chipped the card in the process. With a screwdriver blade, we peeled the dried glue and pressed it back on the floor again. This time with the cameras rolling, the credit card easily lifted the edge and the glue was removed. Just like they say on TV, "See how easy it is?" Sure, after you remove it with dynamite and reattach it.

Chapter 13

Specialized Shooting Challenges: *Shooting People*

#118 How to Shoot and Work with People

Although it sounds nastier than it really is, shooting people just takes a little decorum and knowing when to keep your mouth shut (that works with a lot of things).

This lucky chapter discusses shooting babies, children, and senior citizens. The chapter will also cover two groups that aren't difficult to shoot, but may be hard to get to: sports figures and celebrities. It will also discuss the task of controlling and shooting a crowd of people, as well as the problem of having too many bosses (directors) on a shoot. It's a good thing this type of shooting doesn't involve a gun.

#119 How to Shoot and Work with Babies

Usually, the smaller they are, the more stuff they have in tow and the more difficult they are to work with. Babies are cute, they smell great (some of the time), and they have the attention span —of a flea.

Obviously, when you are shooting an infant, one or both of the parents are on the set. They can help make goofy faces, calm jangled nerves, and feed the budding star every three minutes. I'll give you an example of a typical baby shoot in which the infant's video-savvy mom calls all the shots.

The mother was more than ready for her Lilliputian daughter to begin her illustrious television career, but sweet pea had other things in mind. The action was so simple even a child could understand. The little bundle of joy would be seated on a fine piece of leather furniture as Mommy walked by. This was the client's concept, so we had to appease. The camera would glide down as her mother walked past, admiring the great selection of used-to-be-cow furniture.

As soon as the camera began recording, Baby would look directly at Mommy, who'd be standing next to the camera. This is all the child ever focused on in her life because she was in a papoose and was carried 29 hours a day. With your mother three inches from your face constantly, you would normally be used to watching her every move.

As Mommy walked with the descending jib arm following her movement, so would Thumbelina's eyes. Takes five, six, and seven were no better. Mommy decided that maybe she should stand somewhere else. When the applause died down, the camera rolled again. As Model Mommy held Tiny Tina near the $4,000 leather sectional, the baby decided to mark her territory. A fine mixture of strained peas, pureed spinach, chocolate pudding, and apple juice was deposited on the sectional. Who would have thought that a jostled child would spit up on the furniture?

As the client came running to save the leather, one of the crew suggested that this might be a good way to show the resiliency of the leather. In the real world, kids will end up spilling things on the furniture. Through his tears, the client said that it was a good suggestion but not for this particular spot. Being the director, it was my job to hold the kid as Mommy siphoned up the mess. I had the baby in my arms for 4.7 seconds when she decided to release some more on me. Since she had already regurgitated from the mouth, she tried something new. Of course, this is why God made diapers, but in order for them to work, the clothed infant must be wearing them. Why wasn't she? The child never acted again.

Whenever babies are on the set, everything must be tailored to their fussy schedules. Their time under the lights must be brief and they need a million colorful things to look at for a distraction. Their wardrobe will be changed constantly, and food will be served at the correct temperature, consistency, and color in the appropriate apparatus. Loud noises are inexcusable and everyone on the set will lower their mentality to that of the baby. The set also cannot be too hot or cold (Goldilocks Syndrome), and their on-camera lines must be kept to a minimum.

Babies are fun to work with if you have the time, desire, budget, and the right talent. This is one case where the best director in the world won't be in charge on the set. Only Mom and the baby have that right.

#120 How to Shoot and Work with Children

The horror stories you've heard about child actors may not always be true. I've worked with my share of tomorrow's leaders and found almost all of them to be excellent, professional, and easier to work with than pets. On a recent national spot, I worked with three of the most talented short people under the age of nine in my illustrious career (okay, their parents made me write that). So how do you work with young children in front of the camera without them kicking you in the shin, calling you a meany, or throwing a screaming tantrum that shatters eardrums?

Three gifted eight-year-old girls (sounds like a rock band) were cast as ballerinas in a 60-second spot for a mobility company that manufactures and sells battery-operated wheelchairs. In the opening shot, the three minuscule actors were performing on stage and when the curtain closed, one of them would run to her grandmother waiting in the wings.

As she moved the joystick on her chair, she met her granddaughter midway and hugged her. This wasn't an extremely difficult action to shoot, but it did involve timing on everyone's part, especially hitting the marks with a motorized device.

The chosen ballerinas from the audition were outstanding. These capable kids were prompt, professional, and a pleasure to work with. The young girls arrived early with their parents and went into makeup. Because we were shooting at a high school, the students set up a green room (actually painted green) backstage. These little kids were thrilled to be wearing makeup applied by a professional. If they had tried that at home, their parents may have banished Barney for a week.

In makeup and tutus, the wee performers lined up for their first rehearsal. Kids should never be talked down to on the set (the way I am used to being treated). They should be treated as equals rather than children. They have been hired to perform their roles and will do so if allowed to. Everything the crew is doing should be explained so they know the reason for your madness. When the *assistant camera* (AC) sticks a light meter in their face, they should be told what it does (the meter, not the AC).

It's nice working on a set where you're taller than the talent. I always knew I was being looked up to (I had lifts in my sneakers). The camera was placed on a jib so the performers could be shot from their eye level. As the camera swung in for a close-up, one of the girls didn't expect to see her reflection in the lens. Immediately, she froze and began giggling. This is why it is imperative to rehearse everything with the camera movements before film or tape is rolled. The director should even call "action" during the rehearsals so the talent is used to hearing the terms ("beautiful baby" and "luv ya doll" may be dispensed with).

Mistakes should not be pointed out in front of the group. Any faux pas may easily be corrected by doing something over again. Every crewmember is focused on his or her role and the kids are thrilled to be part of a national spot. You still have to let them have fun and be kids, but the job at hand must be completed. During the course of the day, the kids will need to have their make-up touched up. Frequent breaks keep the kids more energized and allow the crew to nap if desired.

If several people are on the set (231 clients), make sure the kids only take direction from the person in charge (not the parents in this case). If too many people are telling the kids what to do, they easily will become confused (just like me). Anyone with a brilliant suggestion should run it by the director before the kids are just told to do it. This is the only time someone will listen to me on a shoot—because I'm a grownup (a short one, but still one nevertheless).

Kids actually enjoy doing the same thing over and over and will not complain after 50 takes of the same shot. They can watch the same movie 43 times back to back, so repetition is never a problem with them. The lighting should also be diffused; tiny actors may not enjoy a 2K blaring on them, and any light will zap their energy. Mostly, kids are thrilled when they see a piece of Tough Spun begin to smolder (you have to keep them happy and their minds occupied).

I'll never kid around (pun intended) with children on the set because it soon becomes silly. Once I get silly, I have to take my football and go home. Luckily, I think I know when to turn it on or off, and it circumstances like this, it's best to play it straight.

The parents will be of no help unless their protégés aren't focused (nobody likes a blurry kid). If they are distracted, the parent can help bring them back to reality. By the way, make sure you have the parents sign release forms before the camera begins to record. Just have fun working with the kids and try to look at the experience from their level. If that doesn't work, take your crayons and go home. Figure 13-1 shows children on the set.

#121 How to Shoot and Work with Senior Citizens

This topic ranks second in difficulty to shooting children. Older people offer a few more challenges in shooting than the average subject, but the trouble lies in the art of making them look younger. Speaking of lies, your object is to distort the truth slightly about their age by skimming a few years off.

Not every senior citizen demands to look like a teenager on camera, but some people think that video doesn't do justice to their semi-aged beauty. The circuitry in some digital cameras actually softens wrinkles and smoothes out the talent's skin. I know several news anchors who rely on this camera trickery to make them look more appealing. But if the camera can't do the job alone, you must step in and save the day.

Figure 13-1

Working with kids and living to tell about it

When lighting, definitely try a softer, more flattering approach. Hard, directional light will accentuate the folds in their skin, and unless that is your motive, a softer, more diffused light will help. I also add a little warmth to the shot by lowering the color temperature slightly via an amber gel or color balancing on a pale blue card. Once again, orange isn't the goal; soft and warm skin tones are. This slight color addition makes the women look more grandmotherly and makes you want to ask the guys out for a game of catch with your kids.

Besides softening the light, using filters on the camera will also blur the passage of time in their faces. I find using a black net stocking will soften any hard edges and retain their natural skin tone. Other filters like Pro-Mist, Diffusion, and Contrast will all do the same job, with the stocking approach being the easiest to use. If you decide on the glass filters, use the lowest diffusion you can find. You want to blur slightly, not erase.

With age comes wisdom and your talent has worked hard for each and every line on their faces. Don't destroy that by airbrushing out all those character lines. Instead soften them slightly and let their inner beauty speak for itself. Figure 13-2 shows a senior who looks great on camera (I didn't do the lighting).

Figure 13-2

Seniors on camera

#122 How to Shoot and Work with Sports Figures

This section will delve into the arena of shooting sports figures. I've done my share of shooting CEOs (only when they are in season), but I've only videotaped a few sports figures.

I had the opportunity to shoot a video interview with Joe Gibbs, head coach of the Washington Redskins from 1981 to 1992 and current owner of Joe Gibbs Racing, one of the country's best NASCAR teams. Before I left the confines of Pennsylvania and traveled to the friendly state of North Carolina for the shoot, I was wondering why every sports fan I knew was so nice to me. If I had been shooting a celebrity like Mickey Mouse, the age group that was being nice to me would have been younger.

When shooting celebrities, you must realize that their time is extremely limited. (The more important they are, the less time they have available; if I ever get to that point I'll let you know.) On my particular shoot, I had a 10-minute window to get the footage I needed. In order to make this a success, I had to get a number of things accomplished well before the shoot.

Gibbs was to give a presentation at the fiftieth anniversary of a Virginia-based company. Unable to attend in person, he would give a video presentation instead (unlike Al Pacino's computer-simulated *Simone*, he's real). Although his speech wasn't scripted, he had to touch upon certain highlights. If he had been reading his speech, a teleprompter would have been the best choice. But with just key points, cue cards are still the unsurpassed method.

Always prepare the spoken material well in advance of the shoot. I'm sure no one has ever run down to the local office supply store and made the cue cards the morning of the shoot (that wasn't me, just someone who looked like me).

I arrived at Joe Gibbs Racing at 8:30 A.M. in preparation of my 10:30 shoot. A Lowel lighting kit made traveling four states and 500 miles much easier. The back wall of their conference room (his backdrop) was jet black and decorated with Super Bowl and racing memorabilia. His assistant, Cindy Mangum, and I spent a few minutes rearranging the background display to appeal to the viewer without distracting from what Mr. Gibbs would be saying. Figure 13-3 shows the equipment as I attempted to brilliantly illuminate the set.

Gibbs was going to sit on the edge of the conference room table, look directly into the camera, and give his motivational talk. The walls of the conference room were white and the overhead lighting was soft and diffused. In order for Gibbs to be distinguishable from the black background (he was wearing a black shirt), I backlit him with a 650-watt Omni, scrimmed with a metal cookie (a metal screening diffusion) and diffused with a Rosco 216. The right barn door was opened to highlight the colorful Super Bowl jacket, Redskins helmet, and gold trophy in the background. Anything glistening in the background couldn't be too overpowering.

His key light from the right was another 650-watt Omni with amber gel applied to warm up his flesh tones and add a little color to his black shirt. The fill was a third 650 bounced into a white umbrella. Since Mr. Gibbs is a friendly, likeable gentleman, his lighting would be soft and flattering (I guess that's why they light me with black light).

The camera would be eight feet away to capture him in a medium shot. It's important to do your homework on your subject before the shoot. Will the CEO gesture wildly or is he or she rather sedate? I knew my subject was very animated and used his hands when telling a story. When framing for this, I could always zoom in for a close-up.

Figure 13-3

The lighting setup for the Gibbs shoot

#122

After reviewing the cue cards, he decided to change a few points. Most people on that level will have someone else make the changes to the text. Always be ready for that to happen and have the necessary tools. I was fortunate in that he wanted to make his own cue cards (if you want one as a gift, call me and I'll give you a one-inch square piece). In Figure 13-4, Gibbs is making his own cue cards.

While focused on his presentation, he didn't seem to be the type of presenter I remembered from his other television appearances. He was soft-spoken, knew what he wanted to say and how to say it, and seemed extremely relaxed. As soon as I rolled tape, it was as if someone had hit a switch. His enthusiastic, bubbly personality came alive and he delivered his speech to the camera flawlessly.

Afterwards I was given a tour of the racing facility and shot valuable B-roll footage of the cleanest garage on the planet. Not one speck of grease, oil, or dust was found anywhere. I also saw that they park tractor trailers inside (try to get your wife to allow you to do that), have three NASCAR-winning race cars sitting in the lobby (I had to unload all my equipment outside, yet these cars get to be parked inside), and have over 20 cars in various stages of dress, each getting attention from highly trained mechan-

Figure 13-4

If you want the cue cards done right, do them yourself.

Figure 13-5

The completed presentation of Mr. Gibbs

ics (I have to wait two hours at the beauty parlor just to get my haircut). Figure 13-5 is a frame from the video presentation.

It was an extremely enjoyable experience meeting and working with such a down-to-earth, soft-spoken individual. One week later I worked with Richard Simmons, also down-to-earth and soft-spoken.

#123 Review: 10 Tips When Shooting Important People

1. Prepare the script, teleprompter text, or cue cards well in advance. You don't want to be looking for a Sharpie at the last minute. When changes do occur, and they will, smile as you cross out or erase two hours of work.

2. Find out the earliest you can arrive so you can be set up and ready to roll the split second your victim arrives. It's best to be early; if the talent is ready before their set time, you will finish and they can attend to other things. I knew I never should have arrived in my pajamas.

3. Carefully choose a background that won't divert attention from your speaker. I once shot a clown and the elephant in the background was more interesting. Groucho once shot an elephant in his pajamas. "How he got in my pajamas I'll never know."

4. Choose their stance to match their personality or the presentation they will be giving. Some prefer to stand, sit, lean, crouch, or lie down. The edge of a table suggests casual yet powerful, whereas someone standing too stiffly will look uncomfortable. Ask him or her what they would prefer; they'll do what they want anyway.

5. Try to find out what they will be wearing before setting up the lights. A busy pinstripe suit will drive the camera nuts, as will a bright red or white shirt. Light accordingly and if they want to change their clothing after you spent all that time setting up, hit them.

6. Match the lighting to their character. CEOs usually require a harder light to show off their chiseled (not cheap) features. Softer lighting works well on more easy-going, friendly people. I don't usually light differently for men or women and the lights really don't care because they are asexual.

7. Frame your shot loosely until you know what kind of speaker he or she is, keeping gestures in mind. If they hop up on a table and the camera is framed in a close-up, you'll have to tell them nicely to get down.

8. Make sure you get cutaways or B-roll before you leave the scene of the crime. The tape may hiccup, you might kick the tripod, or a busload of illegal aliens (from Mars) might drive by in the background. No www.cutaways.com exists on the Internet as of yet.

9. To keep their time with you as brief as possible (not that you're unlikable), record their rehearsal. Some CEOs are too busy to record a

take two, so don't tell them you recorded the run-through or that's all you may get. Ask them if they would care to do it again. Even if take one was perfect, always try to get another (unless you're driving).

10. If the microphone is attached to the talent, have them hide the mike cable themselves. The last time I went in after a misplaced mike that fell down a shirt, I was in a marriage counselor's office for weeks.

#124 How to Shoot and Work with VIPs (Celebrities)

Of course, everyone you shoot is important, but I'm talking about presidents, CEOs, famous actors, celebrities, or anyone carrying a card proving they are important (if you know where to get one of those cards, let me know).

These types of people usually have very little time for videotaping and you are taking time out of their valuable schedule, even if they *need* you to tape them. You must have your act together before they arrive, have everything set up, and be ready to shoot the second they walk in the door.

Always have cue cards or a teleprompter with highlights of their speech unless they are ad-libbing. Although the on-camera talent may speak for a living, sometimes the presence of a video camera turns people's brains to mush. With highlights or topics listed for them, they will have a crutch to fall back on if needed.

I had the opportunity to work with Richard Simmons at a conference. After hundreds of people listened to lectures all week, Richard was brought in to get their blood flowing again (he didn't scare them by the way he looked; he made them exercise). While the last lecturer was finishing up, Richard was in the background with his entourage entertaining the employees of the convention center.

Being a celebrity, a professional, and a down-to-earth person, he answered every question his followers asked him. Although he whispered, the women surrounding him were cackling with delight at his humor. His groupies, not him, were distracting the last speaker.

Upon completion of the lecture, the master of ceremonies told the mass of humanity that the stagehands needed five minutes to clear the chairs before Richard could take stage. Over 500 stationary, noncollapsible chairs were moved and the ballroom floor was vacated. Richard handed a CD to the audio board operator and told him to play it to get people in the exercise mood. Being an average guy that made it extremely big, he clowned around with us and commented on my beard, the way some of the other crew dressed, and to be prepared for *anything*.

When he took stage, the previous occupant had only about 70 listeners; Richard had hundreds. Stationed on camera one, my job was to frame the stage where Richard was exercising (along with 15 other people) and the people on the floor in a wide shot. My colleague, Greg Ressetar, on camera two had the most difficult job of following Richard as he pranced around onstage in a medium close-up.

The director had given us a basic plan of what to do, but with someone as energetic as Richard, none of us knew what the exercise guru was about to do. As the activity began onstage and each of us set up our shots, Richard would point to someone in the audience and demand that he or she come up and join him onstage. Seeing him point in a wide shot, the director asked camera two to find who he was pointing at and follow them up to the stage.

Luckily, this entire event was only sent live to two large-screen projection TVs at the edges of the stage. Mistakes that happened (there were quite a few) would be forgotten within seconds rather than needing someone to edit the videotape. As we worked our tails off trying to follow the action, Richard would jump off stage and run to hug someone on the floor. Whoever had the best shot captured him while the other operator looked for some type of reaction shot or cutaway.

Ideally, an event like this should have been covered with at least four cameras, two being inadequate with all the jumping around Richard does. But you need to make the best out of the situation and use the tools you're given.

As in any celebrity shoot, you were hired to capture the event to the best of your ability. Few will redo a scene because you messed up. Richard's event was live, so mistakes happened because nothing was scripted, but in our case that added to the madness of the moment. Figure 13-6 shows Richard onstage talking with his guests after the exercise routine.

Unlike other celebrities I've worked with in the past, Richard is extremely approachable. He enjoys when people come up to speak to him. He will actually stay as long as it takes to greet and talk to everyone who wants to meet him. This intimate time should be recorded because it shows the people who made the celebrity who they are.

When preplanning, ask celebrities what they would like you to do. Don't be afraid of them, because they know you are a professional. Most will treat you with respect and I always return the favor. If they should come off with a bad attitude, do not return that same attitude to them. I once had a superstar yelling and demanding that he wanted something; I returned every yell with a calm response. After a short while, he felt like an idiot still yelling. I believe my even tone had a calming effect on him. That situation proved I was more professional that he was (and it kept me from getting fired).

Figure 13-6

Richard Simmons being an average person

Try to defuse any situation before it gets out of hand and enjoy yourself. Opportunities like this don't happen often, so relish them (without spilling any on your shirt).

#125 How to Shoot and Work with Crowds

Whenever you have pedestrians at your shooting location, you will need crowd control. You may not think people will have any interest in what you're doing, but there will be someone lingering who needs to be kept off the set.

I worked on a high-definition video shoot that involved Thomas the Tank Engine. Thomas is a train that has a gray plastic face and is a magnet for children because it's a children's show. Being an adult (physically), I had no clue this train had any appeal to anyone, but I had always noticed its presence being advertised in the newspaper months before it would arrive. Thousands of children in my hometown would bug their parents to take them to see Thomas wherever he might be.

On the day of the shoot, our crew of 12 had to contend with 3,000 children and their parents. As soon as the train arrived, the kids would blindly walk up and want to touch Thomas. They stood in our shots, parents in tow, until their needs were satisfied. I won't go into all the gory details, but keeping people away from a set is a full-time job.

On any shoot, if the public has access to your location, have someone coordinate or corral these wandering nomadic sightseers. If your budget is small and you can't hire someone, contact the police department (if you have already talked to them about a permit) and ask if they can do something to remedy the situation. If they can't help, security companies may be a cheaper alternative.

Only as a last resort have someone on your crew relinquish his responsibility and be a crowd person. Unless you hire him for that specific role, don't pull him from somewhere else because that area will suffer.

If you have any doubts about whether crowd control is needed, check with the police, the chamber of commerce, or the location itself to see what type of crowds they usually attract. Multiply that by 10 to determine the crowd that will show up when they find out you're shooting; a film crew is also a magnet for people (second only to Thomas). Figure 13-7 shows the crew setting up for a Thomas shoot.

#126 How to Work with Multiple Directors

When I worked as a director for a television station, I usually directed the shoots myself (because I was paid to). When I did a furniture store commercial, the owner/manager wanted each shot to have a certain look. In order for him to get that look, he had to be in charge, or direct as I call it. I'm not a tyrant when I direct. I always ask for suggestions and am willing to work with anyone. Not to stereotype, but most muscle-bound, macho-man, tobacco-chewing males don't work as interior decorators—my client did. As he would enter the room, his 6-foot, 5-inch frame would block all the illumination from our lights. As his shadow filled the room like a lunar eclipse, I knew he wasn't happy with something. I could feel a sense of impending doom.

Figure 13-7

The crew ready to shoot Thomas

As he rested his mammoth right arm on his left palm, he stroked his moustache and pondered the situation. Within seconds, his right hand was on my shoulder and he was making a "Tsk-tsk" sound through his teeth. "Something's not right," he said. I knew that and was able to tell him what I thought was wrong with him when I realized he was talking about the shot.

He said that he didn't mind that hot, sweaty, steamy look that would be coming through the showroom if I lowered the light level. If I had realized that he wanted an erotic leather commercial, I would have dressed differently. But since he was the client, I did what was asked of me.

Then came Director Number Two. Our client had hired an ad agency to script and storyboard the commercial. I have worked with many agencies in the past, and we have a great rapport. This particular ad agency guy was invisible. In our initial meetings, he had shown me the 40-color, three-dimensional storyboards on this 12-foot piece of cardboard. Disney had used less detail in *Fantasia*. Besides being one hour late to the shoot, he was never around when we were setting up lights or blocking the shot. But as soon as the camera operator's finger got within one inch of the camera, he would be there. It was as if he had some sixth sense that he shared with dogs. He and the client didn't agree on much. If the client wanted the shot to move, he wanted it static.

As they began to argue about the look of our present shot, Director Number Three came into the scene. It's always nice when directors come out of the woodwork; I have to keep reminding myself to spray more often. This woman had worked as an independent director/producer in the area for numerous years. She had a feel for commercials and was very good at what she did. She had just given birth for the eighth time and had her current sibling surgically attached to her back. As the littlest director would be carried around from scene to scene, she had the nasty habit of reaching out and touching things (she definitely had a future in AT&T commercials).

Every time mommy director walked by a light, baby director reached out and grabbed onto it. As the 60-pound illuminated instrument started to tip, crew members would scramble to their feet to avoid being burned alive. Mommy director would laugh it off and tell Directorette to stop. Every time she would walk past a light, every crew member would be running to stop an HMI from making toast of someone. I suggested gaffer-taping her arms down, but mommy director didn't find any humor in my comment.

If this wasn't enough fun for the day, Director Number Four arrived. Our newest director had been everywhere and done everything on the planet. Although in his early twenties, he told me he had been doing this for 15 years. When he said "doing this," I assumed he meant breathing because that's the only thing he had been old enough to be doing for 15 years.

He began to tell me about the great ideas he had envisioned for this spot. Cameras would be flying around everywhere; he wanted splashes of color and extreme macro shots of fabric detail, but he didn't want the spot to look rushed. As he danced around the room pointing to the images that rushed into his head, I was looking for an aspirin.

When we finally had everything ready for this four-second shot, I wondered which of my colleague directors would be calling action. For the first time, I had every director in the room at the same time. As Director Three, complete with papoose, said "Action," the client said "Stop!" He wanted one of the young people in the scene to have a child in her arms. Where would we get a five-month-old child seconds before we shot the scene? The rental store had loaned the last one to an Asian family.

In a brief moment of madness, Director Three offered her daughter Sweet Pea as the prop kid. Once the strings were cut and Directorette was in the model's arms, action was called again.

Throughout the rest of the afternoon, we directed the spot by committee. It was difficult to find a take that each director liked. If we had been shooting this spot in film as originally suggested, we would have made *Intolerance* look like a short film.

In hindsight, the only way to have avoided "too many chefs" would have been to discuss this arrangement before the day of shooting. Only one person can be in charge on a set; too many bosses causes nothing to get done. Have each “director” in charge of a portion of the spot; what happens in their section they will have total control of.

It may all come down to who is paying for the video. This person is ultimately the boss. But if time is taken *before* the shoot, these problems will be avoided. Never determine this role playing on the set because once you’ve lost control, you will never regain it.

Chapter 14

Specialized Shooting Challenges: *Shooting Animals*

#127 How to Shoot and Work with Animals

There are no two ways about it. Animals are far more difficult to work with than humans. You can try reasoning, but in the end, the animals will win.

In this chapter, the animals are working for us, so keep that in mind. You will learn how to work with our feathered friends, the felines (on a smaller scale), man's best friend (sometimes), and cold-blooded reptiles. You will notice a lot of similarities exist when working with the animal kingdom, but no other group will bite you if you get too close (except the children in the previous chapter).

#128 How to Shoot and Work with Birds

Birds, especially parrots, are very difficult to direct. They seem to do whatever they want to (a lot like my wife). If a bird doesn't want to do something, it won't. The client in our pet shop commercial kept trying to get the bird to squawk and show its colorful plumage. No dice, the bird wasn't in the mood. The client then had a brainstorm.

It seems that the parrot didn't like the jingle used in the spot. Every time the client played the jingle in the store, the parrot would squawk and flap its wings. The jingle was very strange at that. Once you heard it, you'd be singing it all day. The moronic words and catchy tune stayed in your brain indefinitely and made birds scream.

Now that we had our secret weapon unveiled, we were ready to use it on the parrot for a take. For the first time in history since the parrot had heard the jingle, he didn't make a sound. Not even one feather moved on his body. By this time, everyone in the store had been infected by the song. We would be singing it for the rest of the month. If we had had this jingle in World War II, we never would have had to use the atomic bomb.

Parrots are said to be more intelligent that other birds; their brains are bigger (now you know where the term "birdbrain" comes from). Being that smart, the bird would not cooperate with us (I was told later that they only will if they want to). Frustrated, I said that we wouldn't use the parrot, that we would use one of the canaries instead. The parrot heard this and obviously didn't like it; he stuck his black tongue out at me! Luckily, I was able to use this parrot gesture in editing.

You would think that tinier birds like canaries (with smaller brains) would be easier to work with. Whatever our feathered friends normally do, they will cease doing it when the cameras roll. Figure 14-1 shows a canary with a bandaged leg that was more interested in his feathers than the camera.

Figure 14-1

An uncooperative canary

#129 How to Shoot and Work with Cats

One of my favorite animals, the cat, is a lot like a temperamental actor. A cat will do whatever you want it to do at anytime . . . if it feels like it. To make matters worse, my camera operator and I are both allergic to cats.

During a particular shoot, the nonallergic grip brought the cat we would use for the video to the set. The cat immediately saw the fish in the tank we were using as a backdrop behind her. The cat leaped and slammed into the glass. Undaunted by some slight dizziness, the cat started to scale the glass in attempts to stick her paw into the water to retrieve a fish stick. The client must have seen this from across the room because he covered the length of the store in two seconds. That cat's TV days were over.

The stand-in cat was totally different from the bald, hungry one we had just removed. This cat was white, had nine pounds of long, white hair, and the deepest blue eyes I had ever seen. If her name had been Bubbles (and been human), I would have been in love. If we were sneezing with a bald cat, this walking hairball would give us a coronary. As the cat sat in the middle of the set staring at the fish, it did nothing else but look around the room. We weren't that interesting to the feline. As I made bird and fish noises (you'd be surprised what I was willing to stoop to), the cat seemed uninterested. It preferred to stare into the light. Not only did I have an uncooperative animal, but soon I would have a blind one.

As we continued to shoot through our haze of tears and sneezing attacks, we weren't getting much usable footage. As the grip went to retrieve the cat, the cat had something else in mind—me. Knowing that my swollen red eyes and running nose were her fault, she must have wanted to express her sorrow. She leapt onto my chest and clung there for a moment. I knew that I would have a lot of explaining to do when I returned home with 20 tiny puncture wounds in my chest.

As the client extracted the cat from my chest, leaving as much skin remaining as possible, we had our fill of pets for one day. Each animal, like a human, has its own personality and moods. You might have a good day and get an animal to do what you want, when you want it, but often you are just going to have to hope for the best and you probably won't get it.

#130 How to Shoot and Work with Dogs

I'm talking about the quadrupeds, not the unattractive person you may have dated last night. My experience filming dogs includes a video for a new pet store that had just opened in a strip mall that seemed to spring up overnight. Whenever I meet with a client for the first time, I like to come away with at least three different ideas for spots. Each concept will take the video in a different direction, but it's always up to the client to choose the idea they like best.

They say that some actors are difficult to work with, but animals are even more difficult. A pet shop commercial doesn't leave room for as many ideas as some other spots, but I still came up with three totally different concepts. As usual, my favorite one was the one that the client liked the least, and my least favorite was his favorite.

The winning concept involved telling the world about this new pet shop that specialized in dogs, cats, birds, fish, and reptiles. The client chose to tell the world about his new store through a jingle that had been created by a huge agency in Dallas. By using this haunting, silly song, everyone would want to buy a pet from this shop.

When we arrived at his shop on the day of the shoot, we were greeted by a cacophony of animal sounds. Every puppy in the shop was barking, the parrot behind the counter was squawking with an ear-piercing sound (it actually sounded a lot like my boss), and the fish . . . well, I guess the fish weren't really saying that much.

Within a few moments, the decibel level lowered slightly, and we were able to carry on a conversation with the manager. We were to assemble a table as part of the set in one of the narrow pet supply aisles. The aisle was 3 feet wide and 20 feet long. The table was about three feet off the ground and covered with pet carpet. On top of the carpet in the back, we placed an aquarium full of fish. A plywood board partially covered the top of the aquarium; this served as a resting spot for the birdcage.

In creating the animal set, we tried to use as much color as possible to make the area come alive. The carpet was red, the fish were colorful, and every pet accessory on the shelves was handpicked for its bright colors. We also used the standard three-point lighting setup.

Our next task was to find our first victim. We went back to the puppy cages to pick our future star. When I asked for a volunteer, all six of the puppies barked wildly and wagged their tails. It seems as if they each wanted to be the future Lassie, Rin Tin Tin, or Eddie. Their cages were immaculate and it was like trying to choose your favorite candy with only a dime to spend (am I showing my age?)

A small beagle (one of Snoopy's offspring) caught my eye. Although no more than 10 inches long, this beagle had eyes that actually sparkled. And to help her win the audition, she was barking to beat the band (this is cliché week).

Upon opening the cage, the beagle was in my arms within milliseconds. I guess she was glad she was chosen because she continued to show me her gratitude for several minutes. Actually a puppy not peeing on me would have been enough, but I like all the attention I can get.

Once Actress #1 was on the set, we started to roll tape. The puppy just sat there and looked at me. I realized that she was waiting for me to call "Action" (she was a union puppy, Beagles Local 126). I called action and she still sat there. Slowly, her tail started moving; that was the only action I was getting out of her. I quietly asked her to speak, bark, "say anything!" Still the puppy looked at me and said nothing.

The assistant director tried to help by talking baby talk to the puppy in hopes that her squeaking voice would get her to speak. I don't know how babies can stand baby talk, let alone dogs, but somehow when a female talks baby talk, it isn't quite as offensive. Something was happening because the puppy got up on her feet and was beginning to jump in the arms of the assistant director.

By this time, the client had brought over some doggie treats. He thought that if the puppy was chewing, it might pass as movement. The beagle happily took the treats—in one bite. It seems we found a dog that didn't chew. The client returned again, this time with peanut butter. This would stick to the roof of her mouth and I knew some movement had to be recorded. It turns out that our star was worried about her fat intake and didn't like peanut butter.

The client's third attempt proved to be the most effective: dog food. He chose the Alpo brand because it was moist, chewy, and something he sold. Do you know how difficult it is to explain an item listed as "dog food" on your expense statement? This did the trick. As the beagle wolfed down her food, she would occasionally look up at the camera and that would be the take we would use.

The client decided he wanted to use the other puppies as stand-ins in case the beagle didn't work out. When the commercial aired, the pet store received over 200 phone calls about the spot. Each person who called wanted to purchase the talking beagle puppy. If only the caller would have wanted sheep, that would have been a great use for cloning!

The second puppy, a Schnauzer, was just as cooperative as the beagle. Just our luck, we were working with mute puppies. When my neighbors bought dogs, they were never the mute kind we were working with on the commercial. Puppy number two wouldn't eat at all; he wouldn't even open his mouth. I tried making birdcalls and other whistling noises, but all the puppy would do was tilt his head to the left when I made the noises. Cute, but not what I wanted.

The third puppy (a four-pound white ball of fur with a tail) was the most affectionate creature on the planet. If you put anything in front of her, she'd lick it. However, we weren't looking for licking; we were looking for barking. Why did every animal bark at the top of its lungs when we walked into the store, but wouldn't do anything when on camera? Then the thought came to me. Why not have someone new come into the store. That would set the dogs off. Of course, while we were shooting, not one new customer came into the store. Maybe it was the quarantine sign I placed on the door so we wouldn't be disturbed that was causing the trouble.

A few moments later the mailman burst in. What better way to get a dog to bark than to have a man in uniform walk in? Our little ball of fur went nuts. We couldn't shut her up. As we rolled tape during her little tirade, we soon had enough footage of her speaking to fill the Gettysburg Address.

Some closing comments about dogs: They will not do anything you expect them to. If you want a specific action from them, do the opposite, and expect a lot of affection and few other results.

#131 How to Shoot and Work with Reptiles

Although most people go "EEWWW!" when you mention reptiles, we sometimes still must record them on tape. Most reptiles don't move as quickly as other animals, so you may have a little more time when working with them.

I was afforded the opportunity to work with a baby alligator on a recent shoot. All we needed to illustrate was the skin (how ozone depletion makes your skin look like an alligator's). But we thought it might be better to use the entire creature (we couldn't find a zipper on its belly).

We were able to borrow one from a local pet shop and promised to return him the same afternoon when he had finished reading the script. We had water, food (I won't tell you what that was—I would never eat it), and a lovely Styrofoam container where he lived. Because these creatures are cold-blooded, the set must be kept warm at all times because colder temperatures make reptiles sluggish. Of course, keeping an indoor set warm is no problem.

Don't bring your reptilian star out of its trailer until just before you need it. They are more comfortable where they are (in their home) and the disruption will annoy them to say the least. Our tiny, scaly alligator did not like being on camera. In fact, he squeaked the entire time our talent held him. This squeak was the same sound a cat's toy makes when you squeeze it (no, we didn't squeeze the alligator). He was evidently annoyed at my directing ability and was voicing his opinion. Figure 14-2 illustrates our squeak box.

Don't expect any reptile to do the same thing over and over again. If you get what you want the first take, be thankful. It may not happen that way

Figure 14-2

A grinning alligator

again. Try another take, but don't abuse or overwork your future suitcase. Also, stay away from the biting end; those teeth are razor sharp. Respect the reptile and it still won't respect you.

Snakes operate much the same way and prefer to be with someone they are familiar with (a wrangler). Let your snake get accustomed to the set and don't expect it to do something out of the ordinary. Most of the time these reptiles are used for their shock value rather than having them sing and dance. Take your time, have the lighting ready before they arrive, and try to get the shot you're after (before they get after you).

Solutions in this Chapter:

Chapter 15

Specialized Shooting Challenges: *Shooting Food*

#132 How to Shoot Food

This tiny chapter discusses the greatest invention digital filmmakers have when it comes to shooting food: the mirror. This device helps with lighting, shooting, and the placement of your food on the set.

The mirror is an inexpensive way of making a little seem like a lot and professionals have been using them on food shoots for years. Don't you think it's time we did also?

#133 How to Shoot with Mirrors

One of the common problems when trying to shoot food is getting an even illumination into the many cracks and crevasses of the food itself. A carefully placed highlight or shard of light will add appeal to the food and make it that much more enticing. Simply blasting light into the dark spots will solve the problem, but it will also create an overexposed and harshly lit image.

It's now a household word. Everyone uses them at some time, and we are all afraid of breaking them. Yes, boys and girls, I'm referring to the friendly mirror. I'll bet you didn't know that a mirror has literally hundreds of uses besides using it to shave, get at that unsightly blemish, or to apply makeup. One of my favorite uses, that still can be printed, is to use a mirror as a light reflector while shooting tabletops.

Mirrors work equally well in still or motion picture photography. In fact, if you're careful, you can even use them in video. I know you can probably buy a "film" or "video" mirror for $35 from Hammertoes the prop salesman, but you don't need to spend that much. Any Woolworth's (sorry, they're closed), McCrory's (closed too), Kresgee's (I know; it's gone too), or Kmart (that better still be open) has these mirrors at less than $5 apiece. You can even get a two-sided mirror (like some of the people I've known). One side is for normal viewing or lighting, and the reverse side is for magnification. Whatever you do, don't look in the magnification side or you may scare yourself. Unfortunately, the enlarging side doesn't enlarge your light. So if you bounce a 1K into the mirror, you won't end up with a 2K.

As with any reflecting object, a mirror bounces light. Where the light is bounced depends on how the light hits the surface of the mirror and how the mirror is angled. I could get into how the light is actually made up of a series of different wavelengths or talk about prisms and the quality of reflectance, but I won't for the sake of time, and also because I really don't know anything about those topics. I shine a light into it and it works; that's good enough for me.

Now that you know a mirror can be used as a minireflector, how do you use it in a real application? I recently did a commercial where all the stars couldn't talk. Before you call me and ask for their agent's name, I'm talking about food.

The sound stage was a four-foot by five-foot table in the middle of the restaurant. We, the crew, would have to create television magic with hundreds of hungry customers a few tables away. The food would have absolutely no privacy. On all four sides of this centrally located table, we were allowed to place various lighting instruments as well as the camera.

We proceeded to set up a series of makeup mirrors on both the left and right sides of the food. If the tabletop was bigger or we were lighting a larger object, full-length mirrors could have been used. But these makeup mirrors were ideal in that they were small, round, and two-sided; had built-in stands; and will stay exactly where you position them. In addition, we also attached several rectangular mirrors to tiny Matthews stands with fluorescent pink gaffer tape. Some mirrors rested on the table and others sat on 2 × 4 wood blocks.

Once the mirrors had been positioned and I was in a safe place, Inkies and 100-watt Peppers were aimed at the food on the table. At no time did we shine a light directly into the mirror. That would just have reflected the light back to the wall. The trick is to shine the light onto the food, but also have a little of the beam hit one of the mirrors. This way you have control of the light. It's not very often that you have control over anything, so make the best of it.

This reflected mirror light would then shine back onto the food with the same intensity and color balance. If tin foil or a silver or gold reflector had been used, it would have changed the intensity of the light as well as the color temperature. The mirror did exactly what the name implies; it mirrored the light (only backwards, and if you can see backwards light, give me a call).

The mirrors made this particular shoot a breeze. We made great-looking food look even better with slight highlights in the correct places. When people see the finished spot, they probably aren't going to wonder how we lit the food or how it looked so appealing on television. They are just going to rush out to this place and buy exactly what they saw in the commercial (at least that's what I told the client would happen).

Sometimes it's the subtle things on a shoot that make it look really great, and on this shoot I can honestly say, "It's all done with mirrors." That's also a good reflection on the crew.

Small, six-inch-diameter makeup mirrors will permit a tiny bit of light to bathe the subject, as shown in Figure 15-1. Use three or four makeup mirrors in various positions around the food. The overhead Fresnels, softlights, or open-faced units will be picked up in the mirror's reflective surface and bounced onto the food.

Carefully adjust and position each mirror to fill the void created by shadows from the overhead lights. Moving the circular mirror to the right or left, while tilting slightly forward or back, will place the miniscule illumination

Figure 15-1

Mirrors on a tabletop set

exactly where it is needed. The mirrors will only reflect a small amount of light; an accent is sometimes all that's needed.

If a larger amount of light is required, move the mirror in closer or angle its reflective surface. If that still lacks the punch desired, rotate the mirror and use its magnification side. These two-sided mirrors more than pull their own weight on a tabletop.

A mirror will not necessarily give you more light; instead it takes the existing light on the set and moves it to a different location. The only solution for adding more light is to place another light source in an area. Any sized mirror will work, the smaller ones filling in folds and convolutions and the larger ones sending broader highlights.

Chapter 16

Location and Studio Lighting

#134 How to Use Location and Studio Lighting

Before you begin shooting, you are faced with the problem of lighting. In order to expose an image on video, a sufficient amount of light must be present to illuminate the talent or objects involved. If outdoors, will the sun suffice or must you use additional lighting instruments? If inside, how many lights do you need to give you a correctly exposed image? This chapter will discuss your options with both.

What kind of lighting will work best on a set? When on location, you obviously need something that's portable because you have to bring the instruments to your shoot. What type of lighting works best: directional or bounced? If you don't own the right type of lighting, how can you solve that problem? Sometimes building a light that best suits your needs is the only answer.

Outdoor lighting is extremely strong and directional. How do you fight shadows created by the sun? Instead of waiting around for the sun to cooperate, how do you take charge of the situation? How can you use additional lighting outdoors to help when the sun isn't quite enough? This chapter deals with these solutions.

#135 How to Use Bounce Lighting

When blasting light into dark crevasses doesn't work and creates more shine than help, it's time to try bounce lighting. Besides being less harsh, bounce lighting is a more pleasing, softer form of lighting. With only two ways of softening the punch (diffusion and bounce), this style of lighting is preferred because it offers more control. By carefully angling a bounce card (a piece of foam core), you can have illumination exactly where you desire.

Figure 16-1

In a factory, sometimes a bounce is all you can use.

The angle at which the foam core is placed is important. A piece lying on the ground will provide fill light from underneath, and an angled piece will send light in a different direction. The stronger or steeper the angle of the bounce card, the longer the area of light. Of course, light will weaken over a distance, but sometimes that's the look you're after. Take the time to experiment with the position of the bounce card; you can control exactly where the light falls. Figure 16-1 shows an example of an angled bounce.

Point a light a few feet away from a piece of foam core. The light may be open faced or Fresnel; a focusable light will give you more control of the bounce with the ability to spread or narrow its beam. Lean the foam core against your camera's tripod, light stand, or any other object.

You will actually get less light on an object if you bounce an open-faced 2K into your bounce card. However, when using the focusing capability of a 1K DP, the spotted beam of the bounced light actually raises the illumination of the shot. This is one instance where less light bounced actually gives you more. The 2K might give you a broader or wider bounced source, but if you need more light in a given area, a light that can spot its beam will allow you a higher light level. Since more controlled light is pinpointed against the bounce card, it sends more light where you want it.

If your goal is to light a group of shiny objects, surround the area with foam core. A light will bounce around the white area, providing even fill illumination throughout without any of the nasty hot spots. Also remember one thing about bounce lighting. It's doesn't matter what you are using as a bounce source. Only a portion of the light will be sent where you desire; the other part is absorbed. This scientific fact just makes the bounce card hotter to the touch in my opinion.

The exact ratio of bounce versus absorbed light depends on what you're using as a bounce card (white, gray, blue, and so on), but some of the light will never reach the intended subject. That's just the nature of the beast. If you want more of the light to get where it needs to be, you need to use a reflector. A lot of people get this confused.

If you bounce a ball, each bounce is smaller because some of the energy is absorbed in the ground. When you look at your reflection in the mirror, you see all of you (unless it's a small mirror) because the mirror is duplicating or *reflecting* back the same image it sees.

If you point a 1000-watt light against the surface of a white card, you shouldn't expect all the light to fall on its mark; with a *bounce,* it will not. If instead you place a *reflective* surface in front of the light, such as a mirror, a silver or gold reflector, or metallic foil, almost all the light will find its way to the subject, and that's a good reflection on you.

#136 How to Build a Bounce Light

When spending most of your day trying to shoot a ceiling installation video, you need a special lighting setup that will get you through the madness without blowing out the actor's hands, skin, or the ceiling.

Building a bounce light allows you to send soft, even illumination to your white ceiling easier and safer than blasting and diffusing a 2K. George Winchell, one of my favorite lighting people, allowed us to soften the light on the actor's hands, yet still have the ceiling's pattern emerge in the titles themselves.

The unit consisted of an Arri 600-watt Fresnel, but any smaller Fresnel would have done the job adequately. This light was attached to the back extension of a C-stand arm. At the other end of the arm, a knuckle was attached to a three- by four-foot piece of foam core. The arm containing the light at one end and a piece of foam core at the other was attached to a C-stand. This unit could be raised or lowered via the C-stand and pivoted using the arm.

Like the hands of a clock, the bounce light could be positioned from 12 to 11 o'clock, each "hour" giving us a different degree of lighting. Because of the weight of the light at the rear of the arm, the pivot point had to be tight

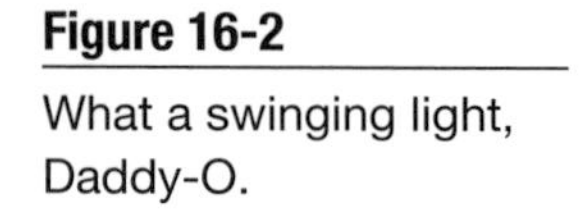

Figure 16-2

What a swinging light, Daddy-O.

#136

or the hands of our clock would swing down to six o'clock at the most inopportune moment.

To keep the light fixed onto the foam core, the unit was barn doored slightly (light blocked slightly) and spotted (focus so its beam was more intense) so all its available light would be hitting the angled piece of white foam core. The unit was situated about two feet from the talent and would act as their key light source. See Figure 16-2 for the only known photograph of our contraption.

If you need more punch in your bounce (sounds like a bad kangaroo joke), substitute a larger Fresnel in place of the 600-watt unit. Keep several of these built units on hand to satisfy your higher lighting needs (like shooting ceilings).

#137 How to Fight Shadows Created by the Sun

What do you do if your shoot involves a remote location far from AC? Since our production company didn't believe in battery-powered lights, we had to work with the sun.

I usually complain about the sun not being bright enough and having to fill in with lighting instruments, but during a particular shoot I had too much sun. Without a cloud in the sky, the talent (demonstrating a new riding lawnmower) was heavily in shadow. No one was going to buy a lawnmower from this shady-looking individual. I have worked on westerns where lighting under a 10-gallon hat was solved, but having no power (something Scotty was always whining about in *Star Trek*) can be an issue.

Luckily, our company *does* believe in Reflectasols. These units have both gold and silver reflective sides. The gold side seems to add warmth, while the silver side supplies much more illumination but is cooler (not a groovy kind of cool, but in color temperature).

Placing the unfolded Reflectasol on the grass and elevating one side slightly, the silver surface blinded the talent and made him squint like Charles Bronson. Now we had too much light. The gold side was much more pleasing and allowed the viewer to see his irises. The completed shot can be seen in Figure 16-3.

Figure 16-3

Filling in light outdoors with a reflector

I guess the moral of the story is, if you have to shoot outside, you really have two affordable options. Use the sun as your friend and bounce or reflect its light where you want it. Or create your own light with gelled tungsten lighting or daylight-balanced HMIs (if you have access to AC). Obviously, each choice creates a different look, but if you're stuck with one type, make it work for you rather than against you. And if you're using lights, don't look directly into them when applying gels or diffusion. You won't be seeing anything else for quite a while.

#138 How to Light Outdoors Without Help from the Sun

Some people prefer shooting outdoors because it's easier, or because the sun acts as the key source of light. But sometimes the sun isn't cooperating with the director and you have to supplement the natural illumination. This isn't difficult and can be achieved in a number of ways.

The main concern after getting an image outside on videotape is color balance. Everyone worth their salt knows that daylight is much colder (or bluer) than tungsten light. If you have a color temperature meter, you would notice that daylight ranges from a low of 4,000 degrees Kelvin to a high of 9,000 degrees. The only reason I know what level a given color temperature has is because our camera displays it in the viewfinder. Video is much more forgiving than film (on color temperature only) and the white balance on the camera will compensate for most colors. I'd like to give you some examples of how we helped the sun along with additional lighting.

Our first problem was when we had to shoot a woman for a tag at the end of a finished video. That doesn't sound like much of a problem because most of you do it every day, but when you ask for the details, that's where the little problems are created. This woman needed a teleprompter for three lines of text. That was no problem. She wanted to be videotaped outside. That wasn't a problem either. We'd just use extension cords to power the teleprompter. The location she chose was under a building's overhang, therefore completely shaded from Mister Sun. Once again, since we had power in the building, we'd just use our lights outdoors. The last thing she mentioned was that she had no budget. That was the problem.

Most people have light kits that travel with them for video production. But like most, the 1000-watt Tota is the strongest light they own. A light that is *that flat* and *broad* doesn't do a great job outside. We were fortunate enough to be able to borrow a 1000-watt DP light from a production house (I agreed to date the owner's daughter when she's old enough). A DP, although a tungsten light, has focus capabilities and can spot its light. That comes in very handy when working outdoors.

Figure 16-4

Using an HMI outdoors to aid in lighting

Since this was our key source, we had to use natural lighting as the fill. The DP was gelled with full blue (raising the color temperature to 5600K). Although the gel cut the light's output drastically, it also softened the illumination on her face. I've rarely ever used a DP without some kind of diffusion; the light is just too harsh.

Our fill was achieved by using a white foam core board. The board caught the natural daylight illumination and bounced it onto the talent. Normally, I would use a backlight or kicker to separate her from the background, but we positioned her just far enough to allow the sun to accomplish that. If the sun wasn't being partially blocked by clouds that day, she would have been too backlit. I don't like to use the sun as a kicker because it's just too powerful. Luckily, the clouds acted as diffusion, so it was only the output of a 250-watt Pepper light (but correctly color balanced).

As Figure 16-4 illustrates, sometimes you need extra lighting outdoors.

Solutions in this Chapter:

Chapter 17

How to Light for Video Conferencing

#139 How to Light for Video Conferencing

We've all been there: sitting in a large room watching a presenter speak, showing slides or videos, and then having to ask the speaker intelligent questions. Now that we've all seen these presentations, how do you light the subject properly without blinding the presenter or washing out the projected images? The key is to make the room work for you (charge at least minimum wage for your time).

This chapter will discuss how to set up a temporary lighting system, one that will disappear after the event. You may be called upon to make an ordinary conference room into a well-lit showplace. Extremely large arenas are also a different beast and will not be discussed here, but a lot of the same basics will apply. But the 1000-person or less event can be lit with a little preplanning and some basic knowledge of what lights will do for you.

In tackling any task or assignment, it is best to break everything down into manageable steps. Since the installation will be in place for a brief time, each step must be checked and rechecked. It's extremely difficult to return to the scene of the crime later and correct a problem that could have been avoided in the first place.

You may be called upon to provide lighting for cameras or just for the audience. It really doesn't matter if the presentation is a one-camera event, a multiunit-switched extravaganza, or a general meeting place for the corporation's masses. The lighting is still one of the most important aspects of the event. I've seen my share of poorly exposed speakers with sunken eyesockets who have to fumble around in the dark, as well as overly lit, glowing stages that no longer have any detail when seen in a video. The ideal situation is a mix between these two extremes.

#140 How to Set Up

It's always best to scout the location well before the first planned shoot or rehearsal. Most corporate presentations of this type rehearse well before the event(s) so the presenters know exactly what to do and when, and the camera crew also has a chance to get their act together. Since this setup will be temporary, knowing what will be needed each particular time is critical to learn at this point. Will the events be recorded on videotape? Will the cameras always be set up in the same positions or should they be flexible? Will the same number of cameras always be needed? Will the presenters always be speaking from the same areas or will that change from presentation to presentation? These questions need to be answered early on before the setup is rigged. Ask any other questions that you feel may be pertinent to your setting up the lighting correctly.

Seeing the place for the first time, we noticed that the conference room was rather small. No more than 50 seats filled the area and it seemed to be a breeze to light. The client then said that the room was currently sectioned off and would be opened for the actual presentation. That meant the room would be six times its present size. Our lighting system would have to work in a sectioned area or the entire space. This little fact made life more difficult. We now had to mount lights that would work in any of the six possible rooms.

Don't just assume the room you walk into will be the same size as during the presentation. It's very common for these larger spaces to be sectioned off into smaller units or expanded to suit the audience's particular needs. Ask the client exactly what size the room will be and how large the audience.

Once the room's dimensions have been determined, discuss the actual events that will occur in these room(s). Ask questions such as how many presenters will there be? Will presenters be at a podium or allowed to pace back and forth? Will a screen be used? Will there be front or rear projection? Ask as many questions as necessary to gain a thorough understanding of the event. Obviously, if you are hired only for the lighting, get all the information you need pertaining to that. If you will be video- or audiotaping the event also, make sure you know what's going on.

#141 How to Light Projection Screens

In our case, this particular series of events featured speakers (12 of them) that would each spend up to 20 minutes at the podium giving their presentation, followed by a brief question and answer session. These meetings would occur every month and would essentially be the same layout, only with the presenters relaying different material each month.

Directly behind the presenters, a four- by five-foot front projection screen displayed the PowerPoint images from the presenter's laptop. In order to properly expose the presenter, care had to be taken not to splash too much light on the projection screen.

Whenever you must light both a person and a screen (front or rear projection), you are faced with some dilemmas. Because you will be illuminating two different reflective surfaces, your lights must be extremely focused, almost spotted. Sometimes it's nice just to throw up a light and expect it to illuminate everything; it will, but you will also have too much spill on the screen, washing out its images. As far as screens are concerned, you *never* need to light them. Their purpose is to reflect light (the PowerPoint or projected images only), and that's exactly what a screen will do with your light source. Your goal is to keep the screen as dark as possible so you won't have an eight-stop difference in light. Contrast will also be a problem, but having too much light is worse.

This is the best time to use a light meter. As a filmmaker, I always have a light meter handy. My meter is an incident/reflective meter. A digital or analog model will work fine. I prefer to use the meter as an incident rather than a reflective meter. By using the sphere on the meter in its incident mode, I can determine the amount of light falling on its surface (the screen, the presenter, and so on). With a reflective meter, it tells me how much light is reflected off of a particular surface. That information isn't as critical as the amount (or lack) of light falling on something.

Walk in front of the screen from side to side, watching your meter. The light reading should be even without any hot spots. If an area is too bright, the lighting will need to be adjusted. When reading a screen, it can never be too dark. The exact opposite is true when lighting the presenter. If the speaker is too dark, the viewer will be watching the screen instead of the person speaking. Take a walk around the rest of the presentation area to make sure the lighting is even.

The first light reading should be taken before the lights are mounted. It's important to see what the ambient light of the room is. The ambient light shouldn't change from show to show, but it's always better to add only the fixtures that are necessary. If you have 200-foot-candles on the presenter with natural room lighting, no need exists to add another 1000-watt light on the grid. Additional light will be necessary, but you may use much less wattage. A foot candle is the amount of light one candle gives off at a distance of one foot.

After you have taken readings of ambient light on the presenter, podium, and screen, make a detailed plan of what additional illumination is needed.

Decide how much light you need on the presenter, podium, and screen. The presenter needs to be the brightest, so the strongest light must be on him or her. It's up to you and the client to determine the ratio of light. Discuss it with the client; should the lighting be high or low key? Will a dimming board be used? Are lighting effects necessary (color gels and so on)?

You also may have a problem when lighting skin tones. In our case, we had 12 different speakers per month. Some of the presenters had very dark skin that absorbed light; others had very light skin that reflected light. A dimming board is really necessary because someone must ride the light level, depending on the presenter's skin tone. Someone with very dark skin and a white shirt will cause a contrast nightmare. You will have to do your best to block the light off their clothing. But with several different presenter heights, you can only do so much.

I usually diffuse every light in addition to using a lighting board. Diffusion like Rosco 216, Tuff Spun, Tough Silk, or Tough Frost works best. Always use the Tough variety because this will soften the blow considerably and is more durable and long lasting. The texture on the presenter's face will change depending on which type of diffusion you use.

When it comes to deilluminating the viewing screen, the only solution is to focus, spot, or flag off any light that may fall onto the screen. This may be accomplished several ways depending on which types of lights you may be using. Lighting may be grouped into three main categories: spotlights, stand- or ceiling-mounted supplemental lights, and practicals. Each type is handled differently.

#142 How to Use Spotlights

The spotlight is just what the name implies. These are usually found in larger presentations that happen onstage. These massive units may be pointed directly at the speaker or fixed onto a grid system. If the talent chooses to move, an operator will have to stay alert and follow the action. Spots can be focused down to a pinpoint or flooded to fill a larger area. Stay with the smallest pattern so the spillage will be minimal. Don't be concerned when using a spot and the talent walks in front of the screen; they will be blocking the image anyway. This is the most controllable and expensive lighting option.

#143 How to Use Stand- or Ceiling-Mounted Lights

Stand- or ceiling-mounted supplemental lighting is what we use most often. Either a lighting grid is suspended from the ceiling (if the room is high enough) or lights may be erected on stands and placed out of harm's way (make sure to tape all the AC cords). Lights on stands may get in the way, but if the ceiling is too low in the room, they may be your only option. These lights may be focusable (if they are Fresnels), but they will not be able to follow the action unless they are the programmable models.

#144 How to Block Light

The best way to get the light only where it's supposed to be is by using barn doors, flags, or black wrap. The barn doors may be opened or closed to position the light only where it's needed. These metal surfaces become hot rather quickly, so use gloves when handling them. You would become hot very quickly too if you stood 3 inches from a 1000-watt light. Have someone maneuver the light as another stands in position with a light meter. You will be able to see and feel when the light has been barn doored off. This takes a considerable amount of time and must be done right. I prefer to get the barn doors in a close approximation while the grid is still on the floor. Once the grid is in position, someone on a ladder may fine-tune each instrument (with gloves).

Flags are much more cumbersome but may assist if barn doors aren't working or if additional light blockage is needed. The drawbacks to flags are that they must be mounted somewhere (using additional stands or clamps) and they are bulky. A smaller flag will block more light if it is held closer to the source, but its edges become less defined. Larger flags are more practical, but in a strong wind they may become airborne.

Black wrap is one of the easiest, most cost-effective ways to block light. The wrap may be attached to almost anything and be crinkled, folded, or bent to the desired shape and size. The only problems with black wrap are the attachment method and heat. You should use clothespins or nonflammable clamps to attach the black wrap. Tape will quickly burn and leave a gooey residue. Although black wrap will block light, it will heat up, smoke, and stink. Luckily, it will never burn, but the last thing an audience member wants to see is something smoldering above their heads.

#145 How to Use Practicals

Practicals, the lighting units that already exist at the site, are obviously the easiest method of lighting. I suggest taking a light reading first with just the ambient, practical lighting. Usually, no elaborate rigs or grids have to be erected; just use the lights that are there.

This does pose some problems, however. Are the lights strong enough (or dimmable)? Can they be rearranged to suit your needs (without offending someone)? Is the color temperature correct (important only if you are recording the event visually) and do you have enough of them? Rarely will the practicals do the job alone. I have only been in one room where that was all that was needed.

Your first meeting with the client is the best time to decide which type of lighting you will use. Don't let it be a surprise to the client. Share your thoughts and see if they will fly. You and the client need to be on the same page, and he or she shouldn't have to worry about your lighting problems.

Solutions in this Chapter:

Chapter 18

Specialized Lighting Challenges: *Lighting Food*

#146 How to Light Food Products

Food has always been one of my favorite things to light because the lighting enhances its appearance. But if food isn't illuminated properly, it looks dull, unappealing, and tasteless.

In this chapter, you'll learn how to light food that you cannot (or shouldn't) eat, food that tastes and looks great (you just have to enhance it slightly), and cleaned-off spuds. The chapter will also cover using chocolate gels to make the greatest food group look its best.

#147 How to Light Edible Food

Let's look at lighting tabletop food. It really doesn't matter which type of food you choose, the same principles apply. Our first example will be a 12-ounce hamburger. On a plate with french fries and a beverage of your client's choice, the food has to look appetizing and edible. While setting up the lighting, use dummy or stunt food. The actual product will look pretty bad after several minutes under hot tungsten lights.

Our key light was a 1000-watt Lowel DP with black wrap on the front. The black wrap was cut into strips so streaks of light would fall across the meat. A 1000-watt Lowel Tota light, heavily diffused with tough spun, acted as our fill light. The backlight was a 2000-watt Mole-Richardson Fresnel shown through a 4 × 6 silk. This silk diffused the backlight so it wasn't a strong source. When lighting food, strong backlighting isn't as effective. People want to see the food clearly, not be distracted by the background.

As I mentioned in Chapter 15, makeup mirrors are essential in reflecting tiny slits of light exactly where you want them. These mirrors are some of the most controllable light-reflecting tools you will find. A small highlight across the front of the burger separates the pieces of ground beef. A handheld compact mirror placed on a C-stand can make the onions stand out, and any other mirror needed can be propped up with a glass, box, or anything else you can find (much like me after a long night of shooting). Our shoot of the hamburger employed six mirrors, each having a specific reflective purpose.

With a Q-Tip, our food stylist used brown food coloring to touch up the surface of the beef. The key to shooting food like this is the depth of field. You don't want the front surface of the meat to be sharp and the rest of it in a blurry haze. A stop of F8 or higher will give you a sufficient depth of field to encompass the entire plate. The drawback with a higher f-stop is that the food will spoil much faster because of the heat. Light the dummy food, and then bring in the real food when you're ready to shoot.

Crab legs seem to always look the same. It's difficult to dress them any other way except making them shine. Our stylist sprayed water on the red legs to enhance their appearance. Garnishing adds color and texture to a plate of red, so yellow lemon rinds and green parsley added the needed color. Our problem was with the small bowl of melted butter. The solution was that the butter was substituted with virgin olive oil (the same color as butter, only thicker).

The client wanted the butter to look more inviting, so trickery had to be employed. A small circular mirror was dropped into our olive oil (pseudo-butter) container. Being lighter than the oil, the mirror surfaced. A small piece of gaffer tape attached to the back of the mirror kept our light reflector from floating to the top. Gaffer tape actually works when submersed in olive oil. Maybe that's where the virgin part comes in. The container now had inner beauty and illumination (much like your significant other).

The same lighting setup, complete with mirrors, was used for the crab legs. Empty film cans elevated our mirrors to make a better reflection on the legs. Figure 18-1 displays one of our tabletop experiences.

The last piece of food to be shot proved to be the most difficult: nachos. When lit incorrectly, a hodgepodge of food like nachos can look like something your pet would rather eat, whereas the viewer would like it to appear appetizing. The only ways of accomplishing this feat are with directional (not soft) lighting and garnishing.

The lighting remained the same as in the other examples, except a splash of color was added to our gray backdrop. A Rosco #44 Middle Rose gel gave the background a festive, Mexican food look. A red linen napkin also added color to the mess we call nachos. After the mirrors had been positioned and the stunt food removed, the real nachos were garnished

Figure 18-1

Tabletop with food, mirrors, and lighting

with shredded cheese (yellow), cubed tomatoes (red), and diced onions (white). Placed on a bed of leafy green lettuce (green), the decorative plate had an international appeal.

The last bit of magic to this spicy entree was the beverage. A small mirror was placed behind a glass of beer. This mirror reflected the key light and made it look as if the glass had been lit from within. Another approach to lighting the beer would be to place a Kino-Flo microfluorescent light behind the glass (a micro flo). This low-voltage, cool light would make the carbon dioxide bubbles radiate. Both approaches work equally well.

I know people who have substituted motor oil for beer (don't drink it). The oil must be clean and a low-weight 5W-30 is the best. Mashed potatoes or sour cream have been called in to act for ice cream, ginger ale has replaced champagne, and lobsters have been polyurethaned in my presence.

With a little imagination and a few mirrors, anyone can make food look better than it does to the naked eye. Just remember one thing: Don't try to eat that pseudo-food that's been sitting under your lights for the last hour. Wait until the actual food has been shot. Trust me, it tastes better.

#148 How to Light Potatoes

Yes, making baked potatoes look good on video is an art. Even mashed potatoes have something lacking when shot without special enhancements (even if Dan Quayle can't spell the word).

Select the best-shaped model for the shoot and clean up its act by scrubbing the spud's skin with water to remove all traces of dirt. To make the skin of a baked potato look nifty, brush it with olive oil. This type of oil will make anything look shiny as well as adhere to the surface. Spray the oil-covered skin with soapy water to create glistening water droplets (don't eat it now).

Ice-cold, raw potatoes can look hot and steamy with audiovisual smoke. Drop some of the A solution SE-502 (it's marked on the bottle A solution) into the open potato with a straw using your finger over the top. The spud will begin smoking when the B solution SE-503 is added. More hot steam is called into action by adding more B to A. If you lose sight of the food through a haze of AV smoke, add slightly less B to the mixture (hard to do if you have already added it). Don't spill any of the smoke solution outside the potato (a steaming countertop is hard to explain).

Nuking the vegetable in a microwave will also generate steam, but that takes a lot longer. Steam can be added to its mashed cousin the same way. Cool Whip stays in place longer than sour cream and semimelted butter will stay that way all day long because the potatoes aren't cooked.

Use every garnish imaginable to add color (chives, bacon bits, and so on). You now have food that looks good enough to eat, but don't.

#149 How and When to Use Chocolate Gel

One of the most exciting things about lights is that you can change what the lighting output looks like. By using things like scrims, silks, or gels, you can actually create exactly the look you want. In this installment, I'd like to talk about using chocolate—chocolate gel, that is.

Just like in the foreign film *Like Water for Chocolate*, a chocolate gel is an excellent addition to your lighting gel collection. This solution will show you the uses and application of chocolate gels, just like the movie did for cooking (without the sex).

If you looked in your swatch book recently, I'm sure you've come across the dark brown gel called chocolate. It is medium brown like the chocolate, but a lot thinner and doesn't smell or taste like the candy. You may also believe that all chocolate gels are created equal, but this is not so. The chocolate gel from Rosco is much different than the one from Lee. For the sake of this story, I'll only be talking about the one from Rosco.

Sometimes you want to add a color to a scene without really changing the color since you may be looking to enhance. You know how popular enhancements are today. Why not do the same thing in video? The chocolate gel isn't really a distinct color (even though it's brown). It's not really a warming color or a cooling color (like *color temperature orange* (CTO) or *color temperature blue* (CTB). Now that you know what it doesn't do, what does it do?

If you're shooting a scene with a gray background, and if you add more light, the gray background will get lighter or whiter in color. If you put a chocolate gel over the light and illuminate the same background, the light only picks up the brown color; the gray wall doesn't get any brighter or lighter. That's because the gel only picks up that particular color of light. It adds richness without really making the scene warmer or colder. You still are actually brightening the background with light because you're using a light, but you're brightening it with color. The gel is raising the light level in the scene without whitening it.

I've also tried this brown gel on wood. I could show you a brilliant example on how this adds richness and an aged patina to any wood background, but you really need a color image to notice this. I rarely shoot anyone with a wood background without using a chocolate gel on my light. You would be surprised at the brilliance (a subtle brilliance) a little chocolate can add.

Here's a wild thought: Why not try to light chocolate with a chocolate gel? It also lightens the chocolate without making it look weird. If you want to add color to a chocolate bar and use an amber gel, it would look too orange. The chocolate gel makes the candy bar look richer. The same physics principle applies; colored gel picks up that particular color and makes it stronger and richer. My Three Musketeers candy bar was so inviting with the addition of a chocolate gel that I had to constantly keep the crew from drooling on it.

Let's try the same gel on a person. Joel Shappiro uses chocolate gel when he's shooting a portrait. By gelling his key light with a chocolate gel and shooting the light through a light black net, it lightens the image but doesn't change the color. This is one of the things a chocolate gel does best. It's also like adding a colored *neutral density* (ND) filter without losing the stop of light.

People with darker skin tones also benefit from a chocolate gel. It seems to brighten the pigments in their skin and give them a healthy glow. Darker objects, skin included, absorb more light than pasty objects. I've had two people standing side by side, one with a fantastic tan and the other looking like a ghoul. The contrast range between the two is almost impossible to light. One appears underexposed and the other several stops overexposed.

To solve this dilemma, use a chocolate gel with the darker-skinned person and either no gel or an amber gel over the lighter-skinned person. In fact, I often just use diffusion on the lighter-skinned victim and a chocolate gel on the other. By doing so, you will have a natural, balanced light between two sharply contrasting objects.

I'm constantly calling Rosco and asking for more chocolate gel. I use mine so frequently that it burns, tears, and often disappears. I believe people think I have an addiction to the gel (if it was the food, some would consider that natural).

Chapter 19

Specialized Lighting Challenges: *Lighting Objects*

Solutions in this Chapter:

#150 How to Light Objects

Objects can be just as challenging to illuminate as food. You have hidden curves that repel reflections, surfaces that absorb light, and some shapes that are almost impossible to light.

In this chapter, we'll talk about lighting bottles, cans, pretty and not pretty hands, body painting (tattoos), products, car interiors, and car exteriors. We'll also discuss how to light smoke, ice, and fog.

#151 How to Light Bottles

One way to make bottles look more appealing on camera is to illuminate them from within. Without resorting to using Plutonium, a light shining through will help just as much (and be less radioactive).

Such a subtle effect may not seem to be worth the time, but every shadow and highlight on the client's product should have a purpose and reason. Just throwing up lights (you really must be sick) and letting the shadows fall where they may is an amateur's approach. Try sending a light stream through a full bottle and see how much it enhances the look. In our setup, we wouldn't be seeing the illuminated bottle, just the colors, shadows, and textures it created as colored shadows on the product.

Bottle number one on our kitchen island setup was a concoction of basil herbal vinegar. This model contained floating debris of rosemary stocks, peppercorns, oregano, and thyme, and it was placed on top of a cookbook (what else) 18 inches from the product (ice cream) and just out of frame. A 200-watt Inkie with

Figure 19-1

Lighting bottles to enhance the mood

a snoot (a cylinder that focuses light on an object) was positioned three inches behind the bottle. By turning the bottle sideways and just missing the label, the light would illuminate all those colored particles in the mixture and make them dance on the product's label. The semi-yellow/lavender glow looked magnificent on the product's multicolored label. At least 15 minutes were spent adjusting this vinegar light just right so every shadow was exactly where we wanted it. It is very difficult to get rosemary and peppercorns to do what you want them to do; I think it has something to do with their past in a salad. But after we got the bottle on the left to cooperate, our attention focused on the more complicated bottle on the right. Figure 19-1 illustrates our complicated setup.

Like its sibling, this bottle of herbal vinegar had all the same goodies inside that the other bottle had, but if you know anything about vinegar, no two bottles are created equal. It would have been fine if this other bottle (there's always another bottle in someone's life) had cast pretty shadows like it should have. We spun the bottle around so much that the peppercorns were starting to homogenize. The shadows just weren't working.

After much frustration, we found that the only way the shadows would work was to have the bottles at a 52-degree angle (no, I didn't have a protractor; I'm just guessing). With a gleam in his eye, our grip reached into his bag of tricks and assembled a unique contraption that would allow the bottle to rest on its side. Two people held the bottle (our prop bottle) while MacGyver clamped a Matthews clamp to its cap. An arm was attached to the clamp, which made its way into a knuckle; this knuckle was then attached

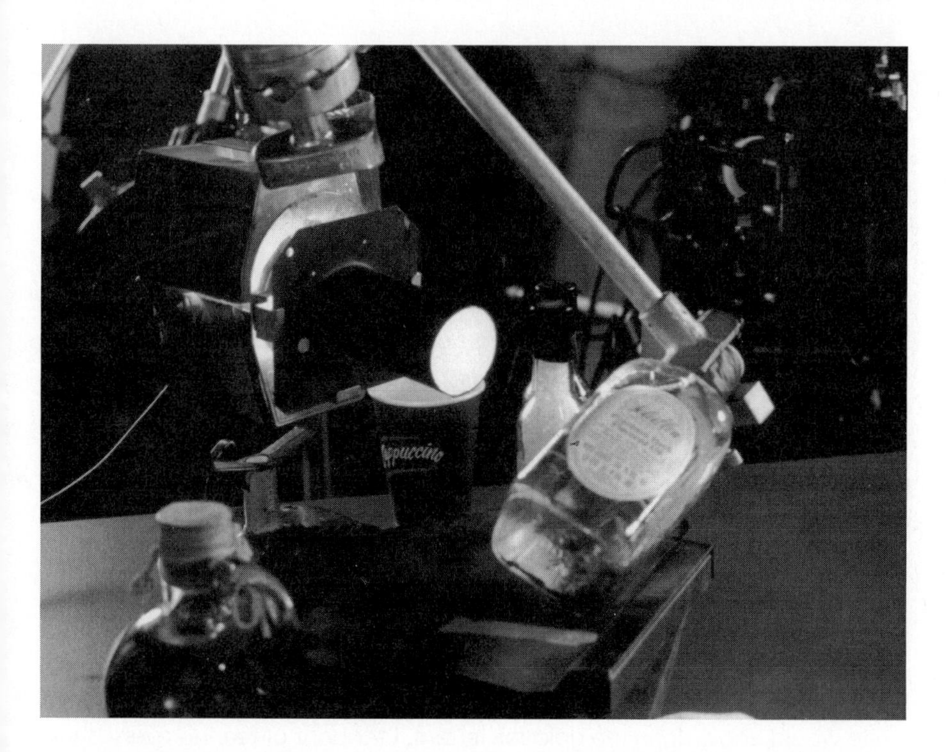

Figure 19-2

Clamp your bottles down so they don't move.

to a C-stand. If anyone breathed, our shadows also would have been lost. Figure 19-2 illustrates the clamping setup. Because of the weight of the bottle and our not really trusting the clamp, two barn doors were positioned under the herbal mix to help take the load off.

We used the same types of lighting instruments, a 200-watt Inkie and snoot, but to add some depth to the shadows, a minus green gel was added to take the green out of the light, as its name implies. This type of gel is used when fluorescent lighting is involved, which normally has a slightly green cast. You might expect a minus green gel to look, well, green. I guess ours was a little too minus because it had a pinkish hue.

This gel took a lot of green out of the vinegar mixture and changed the texture and color of the shadows and highlights. The light appeared far too yellow on the label and gave it a jaundice look. If you are going to use an herbal vinegar in your next label lighting shot, make sure your lighting instrument has a minus green gel and your clamp can support the weight.

#152 How to Light Cans

Have you ever had to videotape a shiny can of polyurethane? On our shoot, the product's name, ZAR, was written on the can in gold foil letters on a solid black label. The camera was to slowly move onto the product as a softly focused brush rested against the can in the blurred background. Some would have taken days to light this product shot; we didn't have that luxury. It seemed nearly impossible to illuminate every letter in the gold, three-letter name. We would get the Z and the R happy, and the A would have a line down the middle. How do you light a round, reflective surface with gold foil lettering?

The easiest way was to basically tent the area with light. Since my Uncle Vinny had borrowed my light tent for a weekend camping (he has a bad back and wanted my light tent instead of my heavy one), we had to come up with another method. I volunteered to flatten the can with my car to ease the roundness problem, but the client wasn't keen on that.

In the house where we were shooting, the camera was placed in the kitchen, shooting out into the foyer. We found that every object in the room was casting its reflection on the can. Upon closer examination, in the reflection I could see my can of Coke on the counter, the grip's jelly donut on its plate, and each piece of foam core we had used to bounce the light into the can.

The only way to block all the unwanted reflections was to make them disappear. If we covered all of the background with something white, only white would show up in the gold foil letters. I thought of tearing every page out of a school notebook that belonged to the home owner's daughter and taping the pages all over the room, but the doodles of the Backstreet Boys' names all had to be erased, which would have been too time consuming.

Everything needed to be covered, including the camera. When we covered the camera, we saw no reflections in the letters on the can, and we also saw no image because the lens was covered. By leaving the lens uncovered and covering everything else in white, we would have our solution. Like Gilligan wanting to help, I pulled every white bedsheet off the beds and came bursting into the room with "the solution." Besides looking like an elopement or laundry ad, my suggestion was rejected. I wasn't referred to as "Blanket Boy." I was just called the boy who was full of sheet.

Then came the real solution: "cone cam." A piece of white poster board (with nothing on it) was wrapped around the lens, but not so you could see it in the viewfinder (Figure 19-3 is the only remaining example of this device). Therefore, everything behind the camera would be invisible, the can reflecting only the white covering. All three letters of ZAR looked great.

In my exuberance, I bumped the table and sent the can rolling on the floor. Our perfect letters would be gone forever. I had visions of spending the rest of my life twisting the can a little to the left or right until all the letters looked perfect. But with two tongue depressors under the back of the can (don't ask why we had tongue depressors on the shoot), the polyurethane was at the correct height. Life again was good.

Figure 19-3

Cone cam complete with white sheets

#153 How to Light Hands

Hands down, one of my least favorite things to light is hands. Did you ever notice how people's hands are always a different shade than their faces? Since we can't change that fact, I'll tell you how I make lighting them tolerable.

I had a shoot at a local nail store (not Home Depot; this was for fingernails). Four Asian sisters and a brother owned the shop. They were true artists and could paint anything on the tip of your nail in extreme detail.

The brother explained that they wanted to produce a 15-second spot for the holidays. People like to decorate their houses around the holidays; why shouldn't they decorate their fingernails? My job was to make the spot as festive as possible, yet still stress the need for people to get their nails done.

They wanted to do several different hands and showcase their art. I told them that it might be more cost effective (that is, cheaper) to show one set of hands from different angles and they would do their best nail job on these model hands. I now had to find a great set of hands (Hand-Mart was closed).

Models are somewhat expensive, but hand models are extremely expensive. The budget for this spot wouldn't allow us to rent someone's hands for the afternoon. I had to use an amateur's.

Just like when I did the podiatrist commercial, attractive people don't always have attractive appendages. I ended up hiring a friend who was attractive and available (did I mention cheap?) for the afternoon.

We tackled this spot on a tabletop. It would be boring just showing fingernails, so we had to have something else going on. Since this was a holiday spot, we thought we'd have the hands wrapping presents. This way the hands would be doing something constructive and also be visible.

A few empty boxes had been wrapped with brightly colored paper and the hands would apply the ribbons as a finishing touch during filming. While the hands were off in makeup having the nails dressed and painted, we lit our tabletop.

Because we were shooting hands and nails, soft lighting was the way to go. By using one 2K softlight as a key, a 600-watt light as a fill, and another as a backlight, we believed we had an evenly lit table. It would be perfect for modeling the hand, and these lighting units provided proper separation between the objects as well. Then when I stuck my hand in to check the lighting, I found out how far off base we really were.

Even with the softlight as a key, the lighting was much too harsh. My hand cast nasty shadows on the wrapped packages. We ended up diffusing the softlight and the others with Tough Frost. The diffusion cut our f-stop, but it also softened the blow.

Checking my hand again under the light showed that the harsh shadows were gone, but my hand still looked pasty (it's not in real life; it just looks that way on video). In order to warm up the pastiness of my hand, I gelled all the lights with amber gel. This gel lowered the color temperature, but gave the hand and the presents a warm, Christmas-like feel.

When the hands (and the rest of the woman) came out of makeup, her nails were blood red with intricate detail painted near the top. The nails looked magnificent. No matter what else was in the shot, the eyes would go to video's least favorite color: red. When designing a shot, the eyes should go to the area you'd like them to; in this case, a vibrant red would be just the thing to do that.

Although her nails looked great, her hands paled in comparison. The model wore rings on all her fingers and I asked that she remove them because they might detract from the freshly painted nails. Little did I know that she was hiding scars beneath the rings. I asked her if she used brass knuckles when she fought. Do you leave the rings on to hide the scars and detract from the nails, or do you remove them and let the viewer look at what they want to: the nails or the scars?

We tried hand makeup, but the scars still came through. We ended up carefully placing the ribbon around as many scars as possible. It looked

Figure 19-4

Scars yesterday, gone tomorrow

somewhat natural, as long as she didn't try to really use the ribbon. If she did, it appeared unnatural on her hand. Figure 19-4 shows how we cleverly hit the scars.

When we instituted a more natural way to drape the ribbon over her fingers, we found out that she had never wrapped a present in her life. I knew she had a boyfriend, so I asked her how she wrapped the presents she gave him. She told me that she always had the store wrap them so she didn't have to. Her secret was now out. If this hit the streets, she would shortly have a new boyfriend.

Here I was ready to shoot a model with wet fingernails and scarred, ribbon-wrapped hands who didn't know how to wrap or apply ribbon. I did the only thing I could in a situation like this. Through a series of fast one-second cuts, we displayed her beautiful nails, had her fondling the ribbon, and still hid her scars. Ah, the magic of a cleverly placed camera, close-ups, and quick cutting.

My advice to those budding hand-shooting videographers out there: Use soft lighting, warm up the shot with gels, and maybe try soft focus or diffusion filters. If you have to hide imperfections, resort to close-ups and quick shots to let the viewers see only want you want them to see.

#154 How to Light Tattoos

Having to shoot a commercial that illustrated the client's business via a tattoo on a young woman's stomach, I believed I have come full circle in this business. After the makeup person had doctored Joy our tattoo model, she was placed on our set. Actually, she walked over there on her own without wrinkling her belly tattoo. Our set was actually in the restaurant and the upstairs area was closed off to the dining public.

A large gray backdrop was used as our background to make our model seem like she was in a void. We would be spending most of our attention focusing on just her stomach so the background might not be seen. A 1000-watt DP with a Rosco #44 middle rose gel created the color cast to the backdrop. A 2000-watt Molette was used as our key light. This open-faced light employs a lamp that illuminates 180 degrees. A sheet of 216 diffusion material was hung between the Molette and our talent. Since the camera would be focusing on the logo, the 216 diffusion shielded Joy from the lamp's light and heat. We didn't want the temporary tattoo to be baked on. The area we were shooting remained unsilked. Figure 19-5 illustrates our setup with Joy.

A 1000-watt Tota light was used as our fill. With two pieces of foam core placed to the left and right of the Tota, the light was bounced between the

Figure 19-5

Lighting tattoos (the lights were hot so she was coolly dressed)

foam core to softly illuminate Joy. The combination of bounced white tungsten light and Joy's alabaster skin made her skin look too gaunt on camera.

George Winchell, our lighting guru, attached pieces of Rosco's Gold Cinefoil to the foam core. The textured gold surface warmed up the tungsten light and made Joy's skin look tanner. An orange gel would have added unnatural warmth to her skin; the gold reflector took the tungsten light and added some slight warmth.

To capture the detail and minute nuances of the logo on Joy's skin, the scene was shot with a high-end digital camera with the zoom set at F11 at 95mm. Unfortunately, while final lighting checks were being made, Joy sat down. Her sitting caused the bull logo to wrinkle. With a second stencil available for emergencies like this, Joy was detattooed, primed, and repainted.

When all the lighting was adjusted, Joy stood in her position. As she raised her arms and slowly turned her waist, inhaled and exhaled, the camera closed into the logo. The subtle camera movement in combination with Joy's movement gave the logo a surreal quality.

When the spot opens, the viewer has no clue that the logo is actually painted on someone's stomach. Later in the spot as more of Joy is revealed, so is the trick.

The next time you're watching television and a logo appears on the screen; try to determine what type of background was used. It could be someone you know.

#155 How to Light Products

As you probably know, lighting helps create the mood. Although few commercials are dark and foreboding, if that's the look you want, light it that way. Most, however, are lit with high key lighting. Everyone's happy and well exposed; nothing's dark and hiding off in a corner somewhere. You should know where every shadow will fall. This is one of the only areas in commercial production where you will have total control.

With almost no exception, the product has to be the best-lit object in the spot. The product is what you're trying to sell, so people should see it in its best light (get it . . . light?).

Under no circumstances do you want a shadow to fall anywhere over the name of the product. People have been shot for less. You need to have control over where the light falls, so be willing to silk, scrim, flag, or do whatever it takes (lighting-wise) to make sure the product is seen.

If the product is to stand out by itself and it isn't with the talent, light the product in its own area. In order to make the product stand out from its

#155

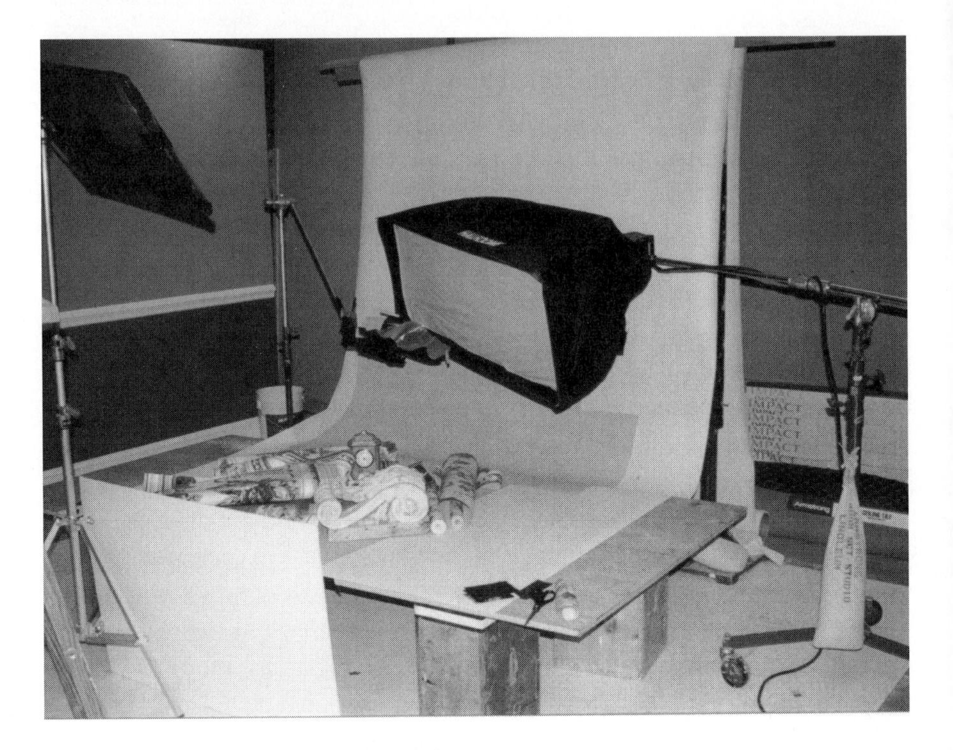

Figure 19-6

A tabletop with a curved cyc

background, drape the background like a cyclorama. Instead of just placing the product on the table and having a piece of background behind it, put it in a void. Pull the background down over the tabletop and place the product in the center. Figure 19-6 illustrates how this is done. With the slight curvature of the background, it will look as if the product is floating in space.

One of the least used and most important lights is the kicker or backlight. This light is used to separate the product from the background. This diffused light should just hit the upper tip of the product, so it is modeled in three dimensions. Without a backlight, the product blends in the background and looks flat and dull. But with a slight highlight, the product will have depth.

The key and fill lights should then be positioned so the product is evenly lit without being too contrasted. This may be the only time people will see the product, and it will never look this good or perfect again. Look at the way food is shot on TV. In real life, the boxes are dented and scratched, the color never looks the same, and the meat is sliding off the bun.

The entire mood of a particular shot can be changed by the angle of the light, the intensity, the color of the gel, and the number of lights. Try to find an f-stop you're comfortable with and increase or decrease the amount of light to achieve that particular stop. Looking at a monitor and saying,

"That's bright enough or it looks too dark," will work in a pinch, but you have control of these instruments, so use them to their fullest capabilities.

To avoid contrast in lighting, I like to employ bounce cards. Somewhat expensive foam core or a cheap piece of white cardboard will reflect or bounce light on the subject. By using gold reflectors instead of silver, a warmer bounce effect can be achieved. It's nice to make living people look like they're living.

Cookies can help define that shadow pattern. I've often cut slits in cardboard to create venetian blind effects, window frames, and other types of patterns that would enhance the shot. Use your imagination. You'd be surprised what you may come up with.

#156 Review: The 10 Most Important Things to Remember When Lighting Products

1. Use lighting to your advantage. Most lights don't have a mind of their own (except on the *X-Files* and that's off the air), so get the most you can out of them.
2. Use diffused lighting whenever possible. Softer light will enhance the look of the subject except hard-water stains.
3. Use gels to change the color temperature. What better way to enhance the mood of the shot?
4. Use a backlight when lighting a product or person. Backlights will separate the objects from the background. The third dimension is here to stay.
5. Use gobos or cookies to create different types of shadows and patterns. Be subtle enough that the viewer will wonder why the shot looks more interesting.
6. Try different types of lighting instruments for different effects. A softlight casts a different pattern than a Fresnel.
7. If you really want a softer look, try using bounce lighting. Point every light at a reflective surface (no bald people, please).
8. Use practical lighting in a scene to your advantage. Make that light on the end table whatever wattage you would like.
9. Make the lighting outdoors work for you. Use lights to fill in the shadows you may not want (use dichromatic filters to keep the correct color temperature). If you use a reflector, don't move with it; viewers catch onto that quickly.
10. Keep things you don't want the viewer to see in shadow. The human eye (and most dogs) will always look at the brightest image.

The key thing to remember here is to experiment. Find out exactly what a light will do for you by moving, focusing, gelling, diffusing, or scrimming it. You can and will create the exact look you're after. I've been in the dark for years. Now that I've been illuminated, I find that it's a very light load to carry.

#157 How to Light Car Interiors

You may not think this would be a problem, but with the advent of tinted glass, gelled windows, and dark fabrics, the interior of a car can be a very uninviting place. The only way to increase the light level in a car is to blast light from the outside or put lights inside. The exterior approach is easier, but you will cause more reflections and lose most of the light through the glass. I prefer to use lights inside the car where I have more control.

Kino-Flos are the only way to go because they output no heat and their illumination is flat, even, and constant. Each fluorescent tube should be daylight balanced (5600 degrees K). This diffused key light will cast a cool glow on frontseat occupants but offer little in the back. These lights, because of their size, should be positioned outside the vehicle. You still will have reflection problems, but fluorescent lights are less complicated than open-faced units or Fresnels. These lights should be your key lights.

The fill lights definitely must be inside the car. Only smaller units work well here. To solve that problem, four Micro-Flos may be attached, two on the front dash and two on the seatbacks via the magic of gaffer tape (if you need to illuminate the backseat). The daylight-balanced Micro-Flos add a sparkle to the occupants' eyes. Powered via dimmers, the Micro-Flos may be cranked to full or half without changing the color temperature (try that with a Fresnel).

The power cables for your lights should run out of the car somewhere to a power source (that's a problem for another day). Figure 19-7 shows some car interior lighting.

#158 How to Light Cars

When lighting cars, you must first position your vehicles where you want them. Now it's time to change your focus to the lighting. If shooting in a showroom, all lights should be silked because of the sea of chrome. In my car, I had to achieve a delicate exposure balance between the vehicles and our on-camera spokesperson.

After using the white Blazer as a bounce vehicle for the lighting, we choreographed the spokeswomen's every move so she could fluidly move between each vehicle and discuss its attributes. Every shine and highlight

Figure 19-7

A car full of lights, and cables. We just need people.

on very vehicle was exactly where I wanted it to be, because I used every flag, silk, scrim, and crewmember in my arsenal.

As an experienced director, when lighting, I always check the circuit breaker box and make sure I'm not exceeding the amperage. Even though I had checked each breaker and none were near their maximum capacity, take 10 began with three of our lights going out. After a bit of sleuthing, it seemed that a bored salesman was playing a full-scale, arcade-type video game in his office. This game should have been running on 220 current. After he was given another task, we resumed our duties.

Check and double-check everything the client says he's done. He may have overlooked something, like telling the talent what to do.

Try to use a vehicle as a reflector. You can even move it if you have to. Although this is an expensive lighting rig, it is sometimes more mobile than a video reflector. Chrome is supposed to shine, so don't worry about the vehicle having highlights and reflections. The angle and curve of the metal will distort the image enough that the viewer won't be able to tell that you used a Chimera light.

Put every shard of light where you desire it most. Once again, cars have reflective surfaces like glass, mirrors, and metallic paint. If you use them to your best advantage, it will be a nice reflection on you. Figure 19-8 shows two NASCAR cars (from Joe Gibbs Racing) that were lit from the outside to simulate natural daylight.

Figure 19-8

NASCAR cars parked inside (I had to park my car outside.)

#159 How to Light Smoke

Working with smoke, ice, and fog isn't too tricky. You can make the almost gas-like substances go somewhat where you want, but lighting it is a whole different story. If you ever tried pointing a light directly into it, all you get is the light reflected right back at you. Fog is lighter and easier to see through; smoke is far thicker and more opaque (like someone I once dated).

Make sure your set has plenty of ventilation. Too much heat or air conditioning will send your clouds to the far corners of the room. Although nontoxic, if humans will be breathing the vapors, have some type of air movement or exchange.

Smoke cookies produce the same effect as fog but in different quantities. Smoke cookies tend to billow and climb, while fog will meander and roll about. Both smoke devices will hiss when the smog escapes, so if you are recording sound, try to anticipate it. A London look works well with fog, and smoke cookies produce walls of white.

#160 How to Light Fog

When using a fogger, a piece of foam core placed above the stream will keep it flowing toward the ground. Fog and smoke have a tendency to rise, so unless something above is pushing it down, everything will collect on the ceiling. By using the foam core as a "fog mover," pulling the card toward you or lifting it up will cause the fog or smoke to rise, while pushing the foam core down will send the billowing mass to the floor.

The fuel in the fog machine determines how much fog is created. This isn't instantaneous, so if you need fog on cue, start the machine before the camera needs to see it. In other words, anticipate the demand.

Lighting the fog from underneath will give it an eerie glow. A soft, diffused light will evenly light the bottom of the fog when placed below and, if above, it looks like snow-capped peaks. Hard directional lighting will send shafts of brightly lit smoke particles in the air. Both looks are quite different; practice to determine which you desire.

Adding a colored gel to the light will obviously give the colorless smoke that hue, with lighter colors being most effective (the *Star Trek* TV series did this all the time).

#161 How to Light Ice

The title of this section sounds like a reference to a new beer. Dry ice, after being dissolved in water, will cling to the floor because it's colder and made of frozen carbon dioxide. If the ground needs to be covered with bubbling, planetary ooze, dry ice may be your answer. The sound dry ice creates is bubbly so if witches or caldrons aren't present, something needs to mask the sound if necessary.

Dry ice, when lit from above, will pool in certain areas. Lighting and gelling from below adds that otherworldly effect. If you need three different layers of this smoldering perfume, try a mixture of all three, because now you know how to control each.

Solutions in this Chapter:

#162
How to Light Places

#163
How to Light Flooring

#164
How to Light Kitchens

#165
How to Light in Confining Spaces

#166
How to Use Window Light with Tungsten Lights

#167
How to Light Outdoors, Softening Large Areas

#168
How to Light Vinyl Flooring

#169
How Not to Come off Half-Baked when Using Cookies

Chapter 20

Specialized Lighting Challenges: *Lighting Places*

#162 How to Light Places

In the last chapter, we discussed lighting objects, and in this chapter we'll talk about lighting entire areas. Whether using daylight or tungsten, or even a mixture of both, you still can control the mood of the area with light.

When you first arrive at your set or location, decide on the "look" you want and begin to assemble your lights to accomplish that. For example, how would you light hardwood or vinyl flooring? How do you illuminate kitchens? How do you light when space is an issue? How do you mix daylight and tungsten without seeing pools of different colored light? How do you achieve soft lighting outdoors? How can you supplement lighting when outside, and how do you use cookies to shape and mold the texture of light? These questions (and more) will be answered in this chapter.

#163 How to Light Flooring

When lighting what's under your feet, most times you will need quite a bit of light if shooting wood. If you just blast light onto the floor, you will have to contend with smears, shines, and hotspots. Wood does absorb light, but illuminating it can be done in a much easier way.

I've shot videos where actors appeared to be working in a black hole. Actually, the floors and walls were mahogany wood and just lighting the actors wasn't enough. Wood surfaces also need lights of their own; they want to be just as important as the talent. All the woodwork in the scene had its own lighting; a light

Figure 20-1

Wood was darker in the 1920s before they had light.

bounced onto every surface. Usually, the wood required a 2K and the actors a 1K (I guess these actors weren't absorbent enough). Figure 20-1 shows how woodwork can look too dark if not lit as well as the actors.

The best way to evenly light a floor is with bounced light. Hard directional light will cast hot, overexposed areas that are unappealing. Take any source of light and point it directly into a piece of foam core. Angle the foam core toward the wood and you now have a great method of soft illumination. The stronger (high wattage) the light you pound into the foam core, the brighter the floor will be (nobody wants a dumb floor).

If the floor is extremely dark and foreboding, you need to introduce more light to raise the footcandle level. A dark floor should not be more than two stops darker than its surroundings. Set up lights and bounce them onto the floor using foam core or the white walls of the room. When your floor lighting level is acceptable, concentrate on lighting the talent.

#164 How to Light Kitchens

It's nice sometimes to have the luxury of working on a set rather than on location. In the tiny world of the set, you are in total control of your destiny.

My favorite part of working on a set is not having to search for AC outlets. One AC cart is more than enough to make every light a happy camper.

We had done numerous kitchen shoots for a particular client and we had a pretty good routine created to light them. This time the client had three kitchens he wanted to videotape for a commercial. Obviously, three kitchens would triple our needed space and require much more lighting. Although once one kitchen was videotaped, we could pull the lights from that set and move it to the next one.

Each separate kitchen was constructed as a set in the studio. Kitchen number one was a light walnut kitchen (if that makes sense), complete with island and separate dining area. Three walls were constructed and the cabinets were attached to the braced and sandbag-supported wall. A matching walnut ceiling was connected to the studio's 20-foot ceiling with wire cabling. The supported ceiling even contained functioning recessed lighting. Ceramic tile was laid on the floor and butted together without the benefit of adhesive or grout. Without examining it closely, the untrained eye would never notice the temporary look.

As usual with most sets that are photographed or videotaped, knobs, finish molding, and other fine touches were left off—only where they will not be seen by the camera. The back of the island that faced the sink was never visible to the camera; therefore, all the knobs, trim, and other pieces weren't attached. This saved a lot of time and was cheaper.

Kitchen number two, the blonde, was much easier to light because of the light-colored wood used. Although the set was just as large as the walnut model, the lighter wood bounced rather than absorbed the light. The walls behind the cabinets were papered and the ceiling was drywall painted white. Crown molding added a nice finishing touch.

Kitchen number three was the real black hole. The cherry cabinet was the most elegant looking, but it absorbed more light than the other two sets combined. Even the wallpaper was dark and sucked up its share of light.

Every shadow on the cherry set had to be created. Because cherry is a dark wood, it absorbs light. Twelve Mole Richardson 750-watt Fresnels were the main punch on the set. Instead of using the overhead grid like most would do on a set (the set's ceiling blocked it), each light was attached to a stand and elevated to the required height. Even with the dark cherry wood, no Fresnel went undiffused. 216, when used in conjunction with the Moles, gave us a focusable, diffused light source. The flooring on this set was tongue and groove hardwood (cherry, of course). This plank flooring was also dry installed without any glue.

Because each set had a fake window, light and a photographic background were used for realism. Figure 20-2 shows how this was done. Mole Richardson 2Ks acted as sunlight when shown through amber gel. These

Figure 20-2

Beautiful vistas on a set

2Ks were the biggest light source on the set. When shined through the window frame (no glass), the shadow of the window would be thrown onto the cabinet surfaces. Usually, at least two 2Ks were needed to pound the light through the empty window slats. By using the barn doors and black foil bent to the correct shape, no shadow was allowed anywhere it wasn't asked to be.

The view through the window was a majestic Colorado Rocky Mountain landscape, complete with a crystal clear lake. Another Mole 750 Fresnel would illuminate this backdrop and add to the realism. I personally prefer the Caribbean, sandy ocean view backdrop, but the client was bigger.

After the first set was lit, it was blocked for camera movement. The client wanted all the camera movement to be fluid, so we rented a dolly and jib to handle the task. The Panther dolly and jib were mounted to a track because the expansion joints in the cement floor caused too many bumps. The pneumatic balloon tires just couldn't handle any imperfections in the floor.

Because every light was performing a specific task, we didn't have to worry about the camera moving through the shadows we created. Larger areas, especially the islands, were diffused with 2 × 4 diffusion stretched in front of several lights. This softer approach to lighting kept the fresh food from spoiling over the many moving camera takes. All the functioning practicals in the illuminated cabinets and ceilings added a bit of realism to the set.

Even though the tile or hardwood plank floors were dry attached, the dolly track had to be shimmed over every joint. It's hard on your knees to stretch a level across the track every few feet, but the end result is an extremely smooth shot. The fluidity of the dolly and jib combination enabled each take to be as smooth as the previous attempt. If one of the flooring joints would separate slightly, it could simply be pushed back together.

On the blonde kitchen, the client decided he wanted to show off the Lazy Susan drawers on the reverse side of the island. Of course, this was the side that had no knobs or trim. Three strong grips simply lifted the stone-and-tile top of the island and spun it 180 degrees. We were now shooting from the opposite angle. Figure 20-3 illustrates how many lights it actually takes to make a kitchen look inviting.

Besides the occasional knob dropping off in the middle of the take (knobs are hot glued on the wood surfaces so no holes will need to be drilled) and a few errant floor tiles sliding under the weight of the dolly and jib, everything went as planned. It's too bad everyone on the crew has to go home to their real circa-1960s kitchens. We all could have definitely made a home out of any of these (even the ones with the plastic food).

#165 How to Light in Confining Spaces

Have you ever had to set up lights, people, and cameras in a closet? For those of you dying to come out of the closet, now's your chance. In order to illuminate this tiny closet, we had to install a light in the crawlspace and work upwards.

Figure 20-3
Lots of lights means a well-lit set (sometimes).

For some reason, green drywall (which is water resistant) was chosen for this closet. Every other drywall surface in the home was white, but this space was green. Somehow a light bouncing off this surface gave the floor an unhealthy green cast. Foam core was taped to the angled surface of the wall, and our light was once again white. In this cramped space, a 750-watt Lowel Omni was used as the light source. Extended on a C-stand arm, the light was placed one and a half feet from the foam core. Figure 20-4 illustrates our confines.

We tucked the light stand as far into the corner as possible. The heat from this unit made the enclosure uncomfortable after a short period of time, but at least we had the light we needed. C-stand arms with foam core can be placed almost anywhere to toss light where it's needed most.

#166 How to Use Window Light with Tungsten Lights

Even the sun is a controllable light source. You just have to let it know who's boss. On a recent shoot, the client wanted his beautiful windows to be seen in the shots of his housing development. If the rooms looked rich, elegant, and filled with love, he'd sell every house. Our job was to make the inside look as appealing as possible. Obviously, you have only two ways to go in the world of video: Gel the lights or gel the windows.

The client had a grand plan. He envisioned a magnificent dining room table festively decorated with the evening summer sun casting long shad-

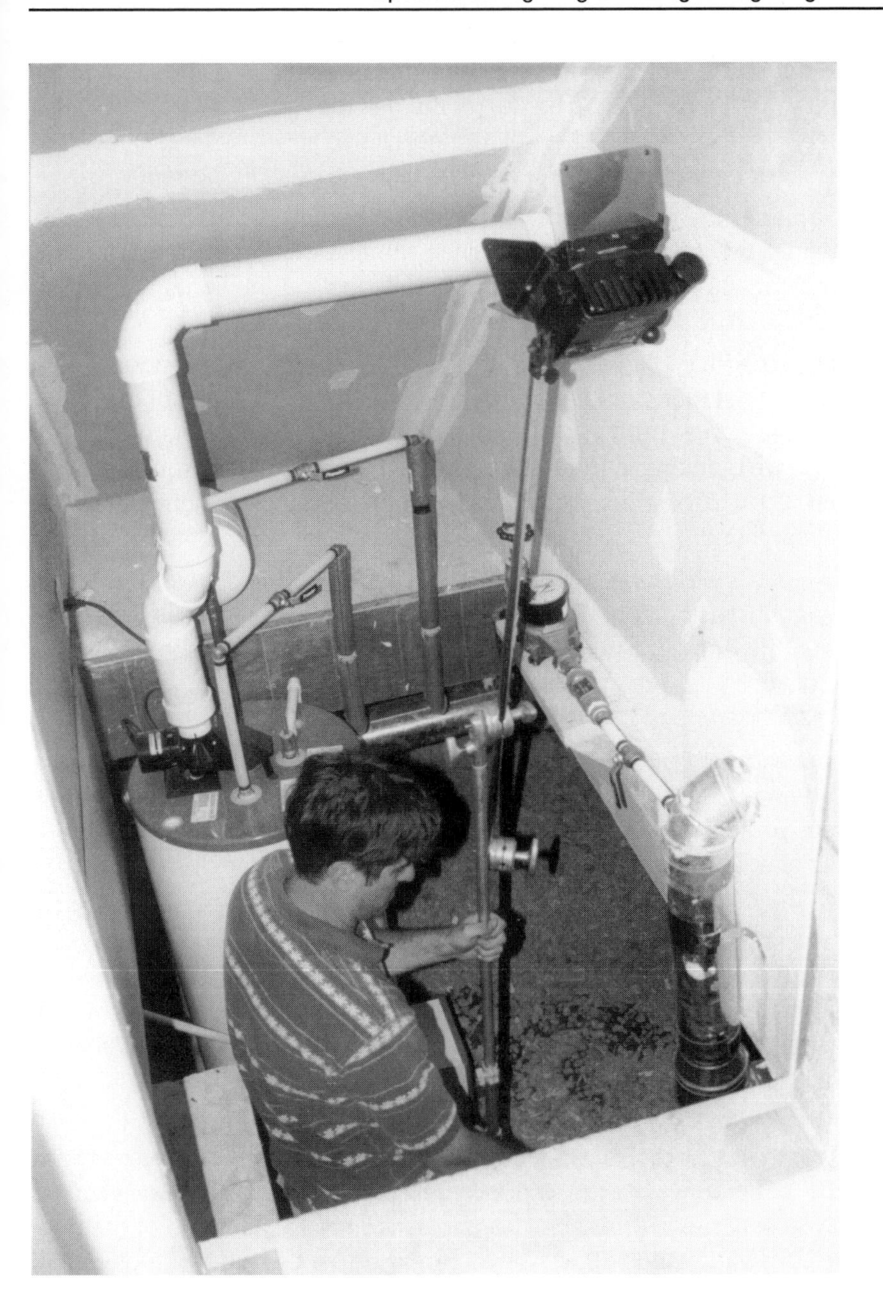

Figure 20-4

Coming out of the lighting closet

ows on the walls. In videoland, achieving an effect like this takes a little planning.

The window in question was the main focal point of the room. Because the dining room was the first room the entering guest would see in the house, the window faced the front porch. Our first thought was to leave the window alone and gel all the tungsten lights.

The drawback with gelling tungsten lights is that you lose quite a bit of light in the gelling process. The darker the gel, the more light you lose. You also need to have enough blue gel. Our gel kit had several pieces of booster blue, but not enough full blue to balance with the daylight correctly.

If we gelled the lights, the light coming in through the front picture window would overpower our lights. Since the client insisted on seeing the individual panes of the double-hung window, gelling the lights and letting the window blow out was out of the question.

We also had a slight problem with power. The house was a model home that had not yet found an owner. The outlets in the downstairs of the townhouse were active, but the units upstairs were not. The house was a gas home, so we didn't have the luxury of tapping into a 220 line from the stove or dryer. Ground fault receptacles in modern homes are a lighting technician's best friend.

Our lights were three 1000-watt DPs, a Chimera with two 1000-watt Totas, and three additional 750-watt Omnis as fill. Since our main source of light was the soft light from the Chimera we could not compete with the power of sunlight. With every light in our arsenal powered up, we still didn't have the amperage to overpower the big guy.

Our only other option left (besides HMIs, Kino-Flos, or Maxi Brutes) was to gel the window and go tungsten. As you remember from your early days in video, you never want to mix color temperatures in a room. Unless the specific effect calls for pools of blue or orange in the room, it's far better to have all the colors talking to each other.

The best way to handle windows is with orange 85 gel and *neutral density* (ND). Sheets of 85ND9 would correctly solve the problem. The 85 would change the daylight color temperature to tungsten and the ND9 gel would lower the light by 2 1/2 stops. With our lights, we now had a more even match with the great outdoors.

Now came the fun part, trying to apply the gel to the window. I usually volunteer for jobs like this when the shoot occurs on the forty-seventh floor and the director wants the window gelled. Luckily, this house was on ground level (I guess that's why no one asked me to take care of the windows). Tom Landis and Brad Kenyon are the two best gel wranglers this side of the Atlantic. They looked at the window and in a flash they were tackling the job.

The gel would be stretched from each side of the frame and gaffer-taped to the vinyl siding. When I looked through the camera, we had a slight problem. Since the entire window was covered with gel, the white overhang (the underside of the porch) could be seen. This white, vinyl-covered obstruction looked extremely dark on the monitor. The solution was to gel only the window as far as the first row on the top of the double-hung window. If the

Figure 20-5

Gelling windows outdoors

camera was kept at eye level or higher, the overhang on the porch blocked all the daylight from coming in the window. All the light that was allowed to sneak in was filtered by the 85ND9. Our plan of attack can be seen in Figure 20-5.

Starting at the window, the camera panned over and tilted down to the table showing the rich, warm sunlight falling in all the right places. It's truly amazing what you can do with gaffer tape and a little orange gel. Just don't let the homeowners know how you did it.

#167 How to Light Outdoors, Softening Large Areas

When shooting outdoors, sometimes the sun is far too callous, blanketing everything with a harsh light. In order to take control of the situation and use the sun to your advantage, it should be diffused. By using a large piece of diffusion material (a 12-foot by 12-foot silk) over the offending area, you can lower your exposure by three stops and have a much softer, more appealing image. Figure 20-6 shows an example of how this tent setup softens the blow.

Hollywood rarely allows the beaming sun to illuminate its sets. Instead, silks are erected over the principle shooting area to soften the lighting considerably. If highlights are needed, additional lights may be placed under the "butterfly" to punch up certain areas. The whole idea of this large silk is to give you more domination of the lighting. You have far more control over the shadows when the main source of lighting is diffused.

Figure 20-6

Under the big top

When shooting in video, watch your exposure latitude. Outside, the object under the silk may be F11. The background on a sunny day might be F32, a three-stop difference. Film will handle this backlight easily, letting the background go slightly overexposed. This same three-stop exposure in video may look washed out. To raise the exposure one stop (a significant advantage in video), use an additional diffused light source on the object underneath the silk. With the subject being at F16 (the f-stop, not the fighter jet), the background's two-stop difference is more palatable.

It is still worth the trouble to use a silk even if you must use additional softlights to raise the exposure. The harsh contrast of direct sun, especially in video, is much harder to control.

#168 How to Light Vinyl Flooring

Ever since color film was created, tungsten and daylight have not been getting along. Both are powerful sources in their own right; each one wants to flex its muscles and stand out as the winner. When lighting a room with a mixture of both, few want a striped blue-orange color look. On a recent shoot, I was able to make both of them get along.

In a tiny kitchen in a brand-new townhouse, a durable new vinyl floor was being installed. Our job was to record this easy installation process and make a tape available for salespeople. What sounded like an easy afternoon turned into a more complicated shoot than anticipated.

The kitchen's walls were an off-white and no overhead lighting fixture had been installed. All the illumination for our camera had to come from another source. The kitchen had two windows: one above the sink and another in the corner. At this point, the only source of light was the daylight streaming in the window. Like most folks, our lighting instruments were all tungsten balanced.

Because the windows would be in every shot, we would have to incorporate them into our lighting scheme. Since the kitchen was on the second floor of the townhouse, we couldn't easily gel the windows from the outside. Although we had sheets of 85ND6 gel on hand, we couldn't open the top window of the double-hung setup (it's manufactured that way). The only way we could gel these windows was to cut the material precisely to length and carefully tape it to the window frame inside, out of the camera's sight. This would be time consuming, we would waste a lot of gel in the process, and we weren't allowed to stand on the floor because the adhesive was setting—that really ruled out gelling the windows. We now believed it would be easier to gel our lights.

Our key light was a Chimera with two 1000-watt Totas. Each Tota was gelled with 1/4 Booster Blue. Because the Chimera softened the light, no additional diffusion was needed. This wheelable key could be rolled into position as soon as the floor dried.

The fill, a 1000-watt DP, was also gelled with 1/4 Booster Blue and bounced off a piece of foam core. This bounce softened the shadows created by the installers. The camera was then sandwiched between these two light sources at the entrance of the kitchen. With our lens, we were able to get the entire kitchen in the wide shot (without setting foot on the floor).

An additional 1000-watt, gelled DP was placed in the corner (once standing on the floor was legal) and bounced off the white ceiling. Our 4,000 watts of tungsten gave us an F4 exposure, with the windows still pounding out an F8. The two extra stops would have been nicely handled with the ND6, but I've already complained about that. Besides, with 2 installers, 3 crew people, 4,000 watts of light, a 10-foot by 20-foot piece of vinyl flooring, sticky adhesive everywhere, and no air conditioning—it was better to open the windows than to gel them.

Another nice thing about shooting installation in a new, unfinished home is that you have plenty of places to hide your lights. Because the kitchen countertops had not been installed, we were able to put our DP light stand in the open cabinet.

Once the shoot began, the camera was white balanced and we received a happy 3400 degrees K. The slightly warmer cast gave our set a more inviting look.

The actual installation of the vinyl was awkward to shoot. Every installer and crewmember (except me) wrangled this massive sheet of vinyl into the kitchen and tucked it between the cabinets. This wall of flooring totally obscured every lighting instrument and we were lucky not to find a DP under the flooring.

The lights behind the camera reflected off the white surface of the flooring and back into the lens. I knew if I had helped, I would have been adhered to the floor.

Once all the cutting had taken place and we had recorded the obligatory close-ups, we were ready to shoot the beauty shot of the installed floor. The client showed up at this time (after we had already squeezed the floor into place) and decided he wanted the sun streaming through the windows and onto his new floor.

Of course, he decided this at 6:00 P.M. when the sun was harshly blaring through our windows. All the camera could see was an overexposed white orb. The sun needed to be higher in the sky to cast a window pattern on the floor (or the kitchen had to be on the first floor). I offered to create a cookie of the window frame, but the client wanted to see the window in the shot as well.

If we opened the window slightly, we could get the sun to shine through the individual panes and onto the floor. Our only recourse was to gel the open window: We had no place to hide a light with a cookie, we couldn't raise a light high enough to shine through our window, and huge, threatening clouds were rolling in.

Our 85ND6 gel was carefully cut and gaffer-taped to the inside of the window frame. If any extra gel was visible, we would cut it to size so we would see no ripples in the material. As the sun continued to play hide and seek with the clouds, we would have a window pattern on the floor, and then nothing.

I started tight on one of the kitchen cabinets and then slowly zoomed out to show the completed floor with the nice window pattern. When the sun cooperated, we had a nice shot. Other times the sun and I weren't in sink (sorry, a bad kitchen video joke).

If you ever run into a situation where you have to mix daylight and tungsten, first determine the easiest method of lighting and then go with the one that makes the most practical sense; it's rarely the easy one. I'm finally over my fear of vinyl after my recovery from the '65 Cadillac hot-plastic-rear-seat-cover-adhering-to-skinny-legs-in-shorts incident.

#169 How Not to Come off Half-Baked when Using Cookies

If you do any kind of lighting, you've probably used your share of cookies. Cookies come in all shapes, sizes, and materials. I've recently come across a few plastic cookies that are inexpensive in the short term; they seem to like to melt. Let's deal with those of you using the more inexpensive, plastic or cardboard cookies (don't feel bad; I'm in the same boat).

If your lighting design calls for a venetian blind effect, I know all of you with budget restraints have purchased a piece of cardboard and cut slats in it. Foam core works better, but it costs more than cardboard. After you have carefully cut the horizontal slats, hold it in front of the light source to see the pattern it throws. It's hard to judge just how long and how wide these slats have to be, however.

It's best to start by making the slats small at first and then enlarging them as the need arises. The old adage in woodworking still applies: Measure twice, cut once. It's very difficult to cut less when you've already cut more. Once you're happy with the size and shape of the cut slats, you must determine if your slat material is going to burst into flames when placed in front of the light for extended periods. If you place the makeshift cookie in front of anything stronger than a 600-watt unit, the cookie won't be long for this world.

One way to keep the cookie from burning, or warping, is to put some diffusion between the light and the cookie. Even if the diffusion is attached to the cookie itself, this will decrease the heat and light intensity enough to get a little more time out of the homemade contraption. The only drawback is that it will lessen the intensity of your shadow effect.

I am amazed at the many uses these cookies have. They can be used right side up or upside down. Cocking the cookie at a slight angle to the light allows the shadow to appear elongated. A light shining directly behind one of these cookies will produce a shadow that looks exactly like the cutout in the cookie. These shapes are usually too weird or defined for subtle images on the wall. You don't want to call attention to these shadows; you just want the viewer to notice them but not realize they are there.

The whole idea of a cookie is to create an effect, but it should be a subliminal one. Unless your shoot calls for little splotchy shapes on the wall, tilt the cookie or move it farther from the light. Better yet, diffuse or gel it so it is less defined.

Even though these new, store-bought cookies won't melt, I always use diffusion behind them to soften the shape and add a piece of single screen in front to further redefine the image. Figure 20-7 should be good to eat; it's full of cookies.

You can further redefine or create new shapes with a worn-out cookie. Simply attach pieces of "stuff" between the cookie and the light. I've used rolled-up pieces of diffusion, clothespins, pieces of wood, metal sticks, fingers, and anything else that might establish a new pattern or shape. Let your conscience and imagination be your guide.

Figure 20-7

Cookies used on a set

Sometimes I've even been bold enough to use a piece of a cookie on a shoot. A little piece of broken wooden cookie can be positioned in front of a light to create its own distinct shadow. I've often used smaller pieces with smaller lighting instruments. A five-inch piece works well when placed in front of an Inkie. Try putting a full cookie next to an Inkie and notice the huge singular shape that appears. A broad light beam needs a broad cookie, and a narrow-patterned light needs a piece. Nothing is wasted in this business if you try hard enough.

Solutions in this Chapter:

Chapter 21

Specialized Lighting Challenges: *Lighting People*

#170 How to Light People

Lighting people is a whole different ball of wax than lighting places or objects. People move, come in different colors and shapes, and will complain about the lights. In the multitude of commercials I've done, I've rarely used an undiffused light. An undiffused 1000-watt light pointed at someone will make them hot, cranky, and washed out. Lights are just too harsh and directional not to be gelled or diffused. If you like the look of a harsh light, try using slight diffusion first.

Light Frost and Tough Silk are great diffusion aids to use without sacrificing too much light loss. My favorite is Tough Spun. Although a much stronger diffuser, it's really the best to use on people (only from this planet). The lighting now becomes textured, softer, and more diffused.

Make sure the light falls exactly where you want it to. This is why God made barn doors. Use them (with gloves) to block off the unwanted light spillage. Flags can also keep areas of the shot free from unwanted light.

Gels are another way to enhance the lighting in a scene. Through lighting I usually like to make people look as warm (color temperature-wise) as possible in spots. By gelling tungsten light with 1/4 or 1/2 CTO, the color temperature is lowered slightly toward the orange end of the spectrum. By using amber, flame, straw, or bastard gels, the shot can also be warmed slightly. This type of color correction is very flattering when shooting women and children. We're still talking gentle color correction; don't make them orange as if you're shooting on the surface of the sun. The cameraman I work with frequently calls this technique (the slight warming of a shot) the "Gloman Glow."

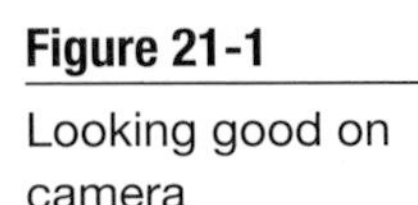

Figure 21-1

Looking good on camera

In the rest of this chapter, we'll discuss if three-point lighting is really necessary, how to light corporate presentations, and how to light with only one unit. Figure 21-1 shows how you can light someone and be very flattering in the process.

#171 Is Three-Point Lighting Necessary?

I should just say yes now and save you the trouble of reading any further. But for those readers who want to know why, let me explain.

The key (good opening joke) is to add dimensionality to your talent or whatever you may be lighting. One light cannot provide any modeling; it just illuminates and creates a shadow. Some crews I've worked with just throw up (I mean set up) a 600-watt light, blast it at the talent, and say the set has been "lit." Yes, it does have more light, but it doesn't look good.

Take the time to set up three lights. The main or key light's purpose is to be the strongest light and do the majority of the illumination. Placed on the right or left, this light should be closest to the talent or object being shot. Place the key high enough so the shadows fall on the ground and out of the camera's view.

The fill should be placed on the opposite side at the same height as the key, but farther away or less intense. This fills in the shadows that the modeling of the key light created. This light will wash away some of the harsh highlights the first light instigated.

The backlight or kicker will separate your subject from the background. Until holographic projection is the norm, he have to simulate three dimensions in a two-dimensional medium (tape). Adding this backlight is the best way to achieve this (and warms the talent's back).

When I teach my students the basics of three-point lighting, I have them light a three-dimensional object and notice what happens when they change the placement of any light. This is difficult to do on the set when the clock is running, but in your spare time, try the same thing. Although I have been lighting for eons (boy, am I tired), I still practice moving the lights around, gelling, diffusing, and scrimming them to see what works best. The Beatles still rehearsed until they disbanded in 1970 and I know you would call them professionals. Follow their lead and practice your craft.

You can always add more lights (hair light, background light, eye light, and so on) if you have the time and equipment, but these three mentioned earlier are mandatory. Figure 21-2 shows how the three-point lighting setup looks to the end viewer.

#172 How to Light Corporate Presentations

We've all been there: sitting in a large room watching a presenter speak, showing slides or videos, and then having to ask intelligent questions. Now that we've all seen these presentations, how do you light them without blinding the presenter or washing out the projected images? The key thing is to make the room work for you (charge at least minimum wage for your time).

It really doesn't matter if the presentation is a one-camera event or a multiunit-switched extravaganza. The lighting is still one of the most important aspects of the shoot. I've seen my share of poorly exposed speakers with sunken eye sockets who have to fumble around in the dark, as well as overly lit, glowing stages that no longer have any detail in the video. The ideal situation is a mix between these two extremes.

Figure 21-2

Three-point lighting

#172

A large pharmaceutical company was having their annual "state of the union" address for their employees and wanted the event recorded on tape. One camera would be stationed at the back of the auditorium to capture the speakers, and other camera (stage left) would allow cutaways of audience reactions.

The featured speakers would each spend up to 20 minutes at the podium giving their presentation, followed by a brief question and answer session. Directly behind the presenters, a four-foot by five-foot front projection screen displayed the PowerPoint images from the presenter's laptop. In order to properly expose the presenter, care had to be taken not to splash too much light on the projection screen.

The images on the projection screen were for the benefit of the live audience. If a camera operator had tried to record these images, a washed out, underexposed, pixilated image would have resulted. The PowerPoint images shown on the screen would then be edited into the finished program using the Media 100.

In addition, the client also wanted the audiences' questions and answers recorded. Therefore, the presenter as well as the entire audience had to be lit to achieve a proper exposure without overpowering the projected image, not an easy task.

Our first attempt at lighting was with an Arri 2K Fresnel. Silked with Tough Spun and barn-doored to a narrow slit, the unit was placed 40 feet from stage left. This diffused and controlled light gave the presenter an F4 exposure without blowing out the projected image. The major drawback was that the presenter couldn't look toward the light without searing his retinas into the back of his skull. Another approach was desperately needed. We needed a properly exposed image for the cameras, and the presenter had to feel comfortable under the hot lights. If the presenter would pass out from the extreme heat, the audience would never learn how the company had done that quarter.

Mike Gorga, the producer, suggested we make better use of the track lighting in the auditorium. These lights, although illuminated, were dimmed to a faint glow. Each halogen unit (15 in total) was positioned 5 feet from the stage area, recessed in the ceiling. With the aid of a stepladder, three of these tiny units were rotated to face the speaker. Because of the height of the lights, the shadows were cast below camera level. The presenter now had a more even, less harsh illumination. White balancing both cameras on the new halogen source created a pleasing exposure without the squinting that the Fresnel created. Figure 21-3 shows the lights we had to work with.

The same approach in lighting was used to illuminate the audience. The additional halogen units in the ceiling were turned from the stage to face the audience. This light allowed the first several rows of people to be exposed at F2.5 and made the audience's view of the screen less washed out.

When the presenter looked out into the audience for questions, he could actually see their faces instead of silhouettes. This practical lighting approach solved several problems. First, no additional lighting instruments were needed for the shoot. Fresnels on stands would have to be taped or

Figure 21-3
Halogen lights in the ceiling

Figure 21-4

The final result with lighting

roped off, taking seats from the audience. Additional 12-gauge AC cables would have to be taped down and run to an adjacent wall for power. Figure 21-4 shows how the room looked to the naked eye. Only what needs to be seen is illuminated; the rest is in the dark.

Second, to avoid harsh shadows, the Fresnel had to be extended as high as possible. The close proximity to the ceiling could have caused scorched ceiling panels, activated the sprinkling system, and create a potential danger to an audience member.

Third, the presenter felt uncomfortable staring at the 2K during rehearsal. A less intense source would be much more comfortable, but when you're lighting a large area, you need the punch a 2K gives.

Fourth, the halogen lights in the ceiling were cooler, less harsh, and rotating them to face the audience removed the wash on the screen that their initial placement created.

Whenever possible, try to use the existing lights in an auditorium. These lights are wired into the house's electrical system and are easier to maneuver than larger units. Obviously, sodium vapor lights shouldn't be used because they are almost impossible to color correct, but most other sources are adaptable to your needs.

When the presenter began speaking, the lights on the audience could be dimmed or raised depending on the situation. Sometimes bringing in additional lighting distracts from the event. Using the natural halogen lighting in the ceiling allowed the camera crew to go unnoticed, resulting in a happy client.

The house's sound system was used to supply our audio needs. The output from their mixer drove our mixer, which in turn went to each camera.

As a first line of defense, try to do what calls less attention to your crew. Most people at these events are there to learn what the speaker has to say, not watch a camera crew in action.

#173 How to Light with Only One Lighting Instrument

It's rare when you find one light that stands out from the rest of the pack. Through years of searching, I have found a lighting unit that is extremely versatile and could almost stand alone on a shoot as the only light source. That light is Mole-Richardson's Molette.

On a recent shoot for the Department of Health in Pennsylvania, the Molette was our only source of illumination used in the high schools. Obviously, we weren't shooting in total darkness in the high schools; the fluorescent lights in the ceiling and several pieces of foam core helped bounce and reflect light, but the Molette was the only additional lighting instrument that was used.

The Molette is a 2000-watt open-faced light that uses a reflector to throw its beam on the desired area. Painted in Mole-Richardson's favorite hue of reddish brown, this high-output lamp acted as our tungsten-balanced sun.

The first step on our shoot involved the physician general of Pennsylvania walking toward the camera. The location was the 50- by 100-foot auditorium/cafeteria in an elementary school. The auditorium's lights were a mixture of sodium vapor and fluorescent 20 feet above the white vinyl floor. Obviously, this type of lighting was a nightmare to white balance.

The key light was the Molette. Pointed at three 2 × 4 pieces of foam core, the Molette's output was bounced off the foam core and onto our talent. Our camera was white balanced, with the talent holding a white card. The strength of the Molette overpowered the other lights in the room. As the key speaker in the video, the talent's lighting had to be warm and inviting; the lighting behind him would be slightly cooler and darker.

The look of digital video is rather cold and hard. As the producer, I utilized a Tiffen Series 9 Pro-Mist filter to soften the hard edges of the digital image. The result was a softer, more pleasing, film-like look. I've been shooting video for 23 years and this Pro-Mist filter is amazing. It takes the edge off the image and really changes the look. If you are after something different in your videos, try the Pro-Mist. You will be pleasantly surprised.

More C-stands were used to wrangle pieces of foam core than were used for lighting instruments. The Molette gave us an F4 and didn't bake the talent in the process.

The next setup involved the talent walking down the hallway talking about the dangers of school violence. The normal illumination in the hallway was a row of fluorescent lights. As he walked toward the camera, he would move into and out of pools of cool light. Once again, the Molette acted as our key light.

If pointed directly at the subject, the bare-bulb Molette tends to cast a harsh and defined shadow. Two thousand watts of raw light will do that with any subject. To soften the blow, the Molette was again bounced off pieces of foam core. Foam core and the Molette seem to be the perfect combination.

The Molette and foam core created another softly lit F4 scene. The hallway behind the talent was allowed to fall off in darkness slightly, the focus of the shoot being the on-camera spokesperson. One other light, for this scene only, was used as a backlight for the talent. A 1000-watt Lowel DP provided the separation needed. The 6-foot, 5-inch spokesperson had to be constantly wary of the suspended Christmas lights that the school keeps hung all year to provide a colorful hallway presence.

Nine C-stands were employed to hold flags, nets, and pieces of foam core. Every shadow in the hallway scene was controlled. With a broad source like the Molette, the lighting will spill onto all surfaces. The dark brick in the hallway absorbed most of the light's output, but the white ceiling and fluorescents provided the much needed fill light.

The last interior scene involved the talent sitting on a table in the school's massive library. Once again, the Molette was the key light source. Angled and bounced off a piece of foam core, he was bathed in a soft, warm tungsten glow. The doors at the perimeter of the room were allowed to blow out. This cold F16 daylight created a contrast between the warmth of the interior and the cold violence that often lurks on the school yard. This indoor/outdoor nonviolence/violence look greatly enhanced the effectiveness of the scene.

A black flag (not the roach killer) was hung above the talent's head as he spoke. This blocked the fluorescent glow that cooled off the warmth we were trying to create. The Pro-Mist filter also caused the fluorescent fixtures to fog and blur slightly. The purpose of the filter was to soften the hard digital edge, but it also added a misty quality to the light. This further enhanced the effect the director, Tom Landis, was after. Violence is definitely a problem in all schools; this dream-like avoidance that most people accept was shattered when the talent went out on the playground.

Once on the playground, the look of the video had to change. This is where most violence starts: outdoors in a group setting. This cold reality had to be emphasized on the video. The long shadows created by an early-winter setting sun provided most of the look we were after. The Molette was also used in this shot. Gelled with Booster Blue, the tungsten color temperature was raised to daylight. Even the Molette can't overpower the sun. Its chief purpose was to soften the shadows on the talent's face; the rest of the playground would be covered with large, foreboding shadows. The actor was meant to be the informative link. If he were allowed to be half in shadow like the rest of the playground, the viewers would not believe he was sincere. The director wanted him to look the same, whether outdoors or inside. He still had to be trusted by the viewer. His surroundings would change, but he still had to be the identifiable link.

The soft look on the talent was achieved again with the Molette and foam core. Flags and nets kept the harsh daylight from destroying the illusion we were painstakingly creating.

Throughout this production, the Molette was our most valuable lighting instrument, possibly because it was the only one. This light was chosen for its versatility and compact size. The last thing we wanted on this shoot was to draw attention to ourselves with an arsenal of lights and kids tripping over the cords. On this shoot we only had one cord: the Molette's.

It definitely is possible to shoot a video with one light (and lots of foam core). By being creative with the foam core, it can be used to bounce or block light. Try to do that with your mother's halogen lamp.

The completed video will be given to all 501 school districts in the state of Pennsylvania. It's my hope that if the public is made more aware of the disturbing situation of school violence, more may be done to put a halt to it.

Chapter 22

Lighting and Shooting Motorcycles and People

#174 How to Light and Shoot Motorcycles and People

Some challenges are involved when shooting motorcycles and people. It's not that the bikes or people are difficult to shoot; it's everything else involved. Whether you are dealing with a location shoot or a closed set, motorcycles are a magnet for all ages of people (those that have and those that want to have motorcycles).

In this chapter, you'll learn from my mistakes: how to light motorcycles outdoors, how to avoid colliding with pedestrians when shooting on the sidewalk, how to avoid losing power in the middle of a take, how to make motorcycle people look happy and convincing, how to deal with cloud cover, how to keep calm take after take, how to use an inverter to change your vehicle's battery from DC to AC, how to get the actors to "do it again, only backwards," how to get everything right in the first take, and how to light a motorcycle indoors when you are in command.

#175 How to Light Motorcycles Outdoors

Motorcycles are fun to shoot because the chrome, high-gloss paint, and throaty exhaust paints an image in people's minds that is hard to dispel. I'd like to discuss how to correctly illuminate a bike outdoors where the only control you have is from your support hose.

The owner of a jewelry store wanted his motorcycle to be featured in all his advertisements. When the client arrived, I could hear his vehicle approaching from a block away. The guttural sound of the engine could be heard above the clatter of the

Figure 22-1

The last known photo of the crew on the "motorcycle sidewalk" shoot

diesel bus engines in our overcrowded downtown location. The bike's sound was pure bass and could be felt on the sidewalk. It was almost as if time was standing still and everyone on the sidewalk started to walk in slow motion. This glistening, mammoth piece of chrome and steel was slowly approaching gently through the carbon monoxide mist.

A six- by six-foot silk was erected over our section of sidewalk to diffuse the scattered sunlight. The sun on the chrome was too blinding for the camera so the silk helped soften the sharp highlights. Blue-gelled tungsten softlights provided the fill to our outdoor setting. Every AC cable was neatly tucked or taped down and we had the permission of the local police to do what we wanted on the sidewalk. Here is where everything fell apart. Figure 22-1 shows the scene of the crime. Every time I called "action," something out of the ordinary would happen.

#176 How to Avoid Collisions

A UPS delivery man ran out of a store and headed back to his truck, never expecting that a motorcycle would be there blocking his way (I guess it wasn't there when he walked in). I learned that no matter who runs into a motorcycle, the bike is going to win. It looked like a classic scene from a *Blondie* movie where Dagwood runs out of the house, crashes into the mailman, and sends letters flying into the air. Lesson learned: Check every doorway and each opening for a potential problem.

#177 How to Avoid Power Losses

In the middle of Take 17, all the lights died. Not only had we kicked a circuit breaker in the store, but our inverter in the van didn't want to work on its own, so it stopped also. The inverter hadn't been used in a while and was getting lazy. The bike was driven into the shot and everything went dark. Lesson learned: Check all your circuits and make sure you aren't pulling too much power. An inverter can only do so much; make sure it's up to the task.

#178 How to Work with Facial Expressions

Because I had so many problems earlier with magically appearing pedestrians, I wasn't paying enough attention to my stars on the bikes. When they would drive by the camera and stop in front of the store, they would be grinning ear to ear, looking directly into the camera each time. It looked a little hokey. I instructed them to look ahead and not into the camera. "How do we know where to stop?" was someone's reply. He had driven the bike to the same spot at least 20 times and now he didn't know where to stop. I placed a small cigarette butt on the mark and that gave him something to look at other than the camera. Lesson learned: Rehearse everything numerous times before the cameras roll. The bike won't foul up, but the people maneuvering it will every time.

#179 How to Work with Clouds

The entire day was overcast, but then on the best take the sun came out and blinded everyone, including the talent. Would you buy jewelry from someone who rode a bike up the sidewalk squinting? Lesson learned: Watch the sun and anticipate what it's going to do. The sun was still our key light, so it pulled the shots.

#180 How to Do Multiple Takes Without Fatigue

We did the same take over and over again. Each time something wouldn't work out just right. I was getting tired, but I never really thought about our talent. Each time he drove the bike into the scene, something wouldn't look right and I'd have him redo it. He complained that backing up a huge bike wasn't as easy as it looked. Lesson learned: Don't practice too much because you will wear out the talent as well as yourself.

#181 How to Use an Inverter

I wasn't aware that when you run a vehicle's inverter, the engine is supposed to be on to continually charge the battery. When the inverter stopped running, so did the vehicle's battery. This kind of makes sense. You are pulling juice from the battery, so how is it supposed to replenish itself? After taking a 10-minute break while we jump-started the battery, we were back in action. Lesson learned: Know how to operate your equipment before the shoot begins.

#182 How to Reverse the Shot

After we had finally completed the opening shot, I instructed our talent to drive away for the closing shot. From the starting point in the front of the store, he would just drive away. Luckily, the closing shot was much easier than the opening shot. In three takes, it was completed. Lesson learned: Save the best for last and people will like you more.

All in all, this was an excellent learning experience. I think I learned more about what not to do on a shoot rather than what to do. I did receive a lot of comments about the spot over the years. Not many people remember the bike being on the sidewalk in the middle of town traffic.

#183 Getting It Right the First Take

A new opening for a biking DVD needed to be shot with on-camera talent Michael "Mad Dog" Ovadia. A 1994 Harley-Davidson Softtail Fat Boy was parked on the edge of a grassy knoll with its gleaming chrome 80-cubic-inch engine facing the camera.

As the sunlight filtered through the mighty towering oaks and the cicadas serenaded us in the 90-degree heat, Mad Dog, clad in black leather, straddled the massive bike and spoke to the camera. Wanting him to appear more dominant in the frame (to some, leather clothing is all you need), I wanted a low-angle shot. The tripod couldn't be lowered enough, so I placed the camera on a cinderblock (I cleaned it first). Five inches above the gravel driveway, the bike and rider combo looked great. Figure 22-2 gives away our secret to an extreme low angle.

A wireless microphone was clipped to Dog's leather vest and the receiver rested with me on the cinderblock (always supervised). The pickup of this lavaliere is nothing short of amazing. If a squirrel coughed in a nearby tree, I could tell what he had for breakfast by the rattle in its lungs. With the mic

Figure 22-2

A boy, his bike, three other boys, and a camera on the ground out of frame

placed very low on Dog's vest and having him speak in a quiet conversational tone, the levels were set. I never get affectionate with microphones, but I certainly came close with this one.

When the Harley was started, the foam in my headphones melted as they rocketed off my head, exploding into the crushed stones. The throaty rumble of the exhaust churned the undigested taco in my stomach as I tried to replace the high-pitched squeal in my ears with normal hearing. I never thought an 80-cubic-inch engine would be 12,000 times as loud as my 250-horsepower car (I guess that's why I'm not as cool as I used to be). With a new sound level set (anticipating the engine noise), we focused on the lighting.

Since the sun was mottled on the talent's face, we blocked the light on everything except the bike's chrome using foam core as flags. Our grip would have to bend and stretch as the constantly changing light would streak the talent's face (luckily that's all that streaked that day).

Mad Dog had to stop the engine, lower the kickstand, take off his sunglasses, lean toward the camera, and deliver his lines. I'm glad he has talent because I would have needed training wheels in order to keep the Harley from falling over on me. Using backlight to our advantage, Dog glowed on the bike. Video can handle backlight as long as your foreground isn't more than four stops different than your background.

#184 How to Light Motorcycles Indoors

After you've mastered the art of shooting a motorcycle under God's light, it's time to do the same thing indoors with a model. How do you accentuate the bike's best features while doing the same with the model? Most models I work with don't have chrome or lacquered body parts; maybe I just haven't worked enough in this field.

I was allowed the use of a home garage and a 2000 FXR hand-built, custom, 110-cubic-inch street bike for our shoot. Since my legs look really bad against chrome, we hired a model to pose on the bike for our shots.

Before she arrived, we had to transform a two-car garage with tools, grease, toxic fluids, and posters into a video studio that would sell our concept. We would have had a lot more control in a studio with a cyc, but our limited budget and timeframe didn't make that an option.

When we arrived at 8:00 in the morning, the temperature was already a heady 85 degrees. With an expected high of 98, the heat index would hover around 112, factoring in the humidity. I wasn't worried that our lights would generate too much heat because women don't sweat (so I've read); they glow, perspire, or wilt.

Before any shoot, you must develop a look (not the kind where you twist up your face and cross your eyes). With this plan in mind, light your set so you have control of the situation. My vision was a brightly colored background that would blur slightly, allowing the warmly lit model to be the center of focus. For the viewers more interested in the jet-black bike, the highly polished chrome had to be accentuated. Our lighting happened to be in kit form. Armed with clothespins, gels, diffusion, and silks, I knew we could get the exact look we were after.

With the bike parked at an angle, we gelled a 1000-watt Tota with rose gel and flooded it against the workbench along the wall. Looking like a brothel on a Wednesday evening in August, within an hour it would smell like one too. On the left we used a 750-watt Omni, also rose-colored, pounding its light on a red Craftsman toolbox. I wanted the background to be three stops less than our foreground subject, which would be exposed at F2.8. With the depth of field being somewhat narrow, the rose would blur slightly (blurry roses cost more).

Whenever lighting a shot, don't rely on the camera's auto-exposure to set your f-stop. Instead, use an incident light meter (which reads the light falling on a object, not reflected off of it like the camera's metering system) to get a more accurate reading. This way you know if you have any hot spots or areas that need more illumination. Remember, anything more than a four-stop range is pushing video's capabilities.

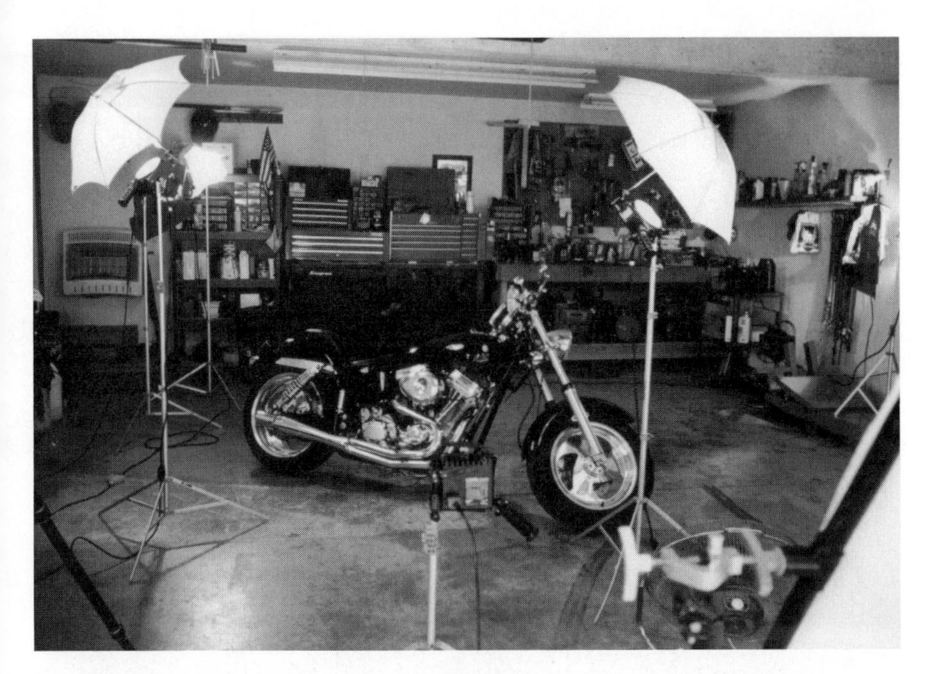

Figure 22-3

Kit lighting (from two kits) and a motorcycle

The opaque black and chrome bike only got one light all its own. An Omni, 750 watts, was scrimmed with a metal cookie and pointed directly at the gleaming chrome engine. The light's barn doors were partially closed to avoid it spilling onto the black lacquer. This light was ungelled because chrome looks fabulous daylight balanced with orange tungsten light. Figure 22-3 shows how kit lighting can do the job.

Reflections of our lights could be seen in the ankle-deep polish of the motorcycle, but the highlights sparkled. You need to be careful when shooting chrome in video because it has a tendency to bloom and overexpose. We lit the bike so the zebras, set at 100 percent, were just showing. Because of the elevated outdoor temperature and our tungsten lights, we left the garage door open, allowing daylight to filter into our set.

Before you start yelling and asking why would I possibly want to mix daylight and tungsten on a video shoot, I had a good reason. I could have kept the door closed and shot with a tungsten balance. But the heat from the lights would have baked us alive and I smell like burnt ham. I also could have balanced all the lights to daylight, but that would have made the shot look too cold. If you want a warm-looking shot, it's best to balance daylight and allow the tungsten to warm up the shot (with a little help).

Another 750-watt Omni silked with Rosco's Tough Spun supplied the model's hair light. Because her hair was a vibrant blonde, I wanted the warmth (orangeness) of the tungsten against our daylight-balanced camera

#184

Figure 22-4

The cover shot achieved with kit lighting and gels

to make her hair appear that much more blonde (we could have called her Blondie, if not for the copyright infringements). If women would just carry around tungsten lights outdoors, they would never have to bleach their hair.

The key and fill lights (on the left and right of the bike) were also 750-watt Omnis bouncing off silver/white umbrellas. A gold-surfaced umbrella would have warmed the shoot too much. This softened the light enough to bathe our talent in diffused illumination. In order to raise the color temperature slightly (toward the blue) and not make everything orange cast, I stuffed Booster Blue gel into the umbrellas. This blue cooled down the shot sufficiently (not that I was getting hot). Figure 22-4 shows the finished result of the shoot.

In order to make our model more enticing (like we needed to), we had our grip hold a fan at the model's eye level (not a small person wanting an autograph, but an electric fan). For the closing shot, we had her wear a black leather vest and stand in front of the bike. With the fan still blowing and shooting from my knees, I had the shot I always wanted to do—shooting a classic bike in a garage in the summer. What did you think I meant?

Chapter 23

TV Commercials

Solutions in this Chapter:

#185 How to Do TV Commercials

Although glamorous to some, the TV commercial is a lot of work and effort for a few seconds of screen time. However, no other creative video outlet allows your work to be seen by more people, broaden your skills in storytelling, and teach you to work under very strict deadlines.

In this chapter, we'll start with a sticky subject: tattoos. From there, we'll move onto how to light a commercial, how to tell and shoot a story effectively in your limited timeframe, how to record sound, how to edit, and how to survive a commercial at a kiddie park.

#186 How to Apply Removable Tattoos

One of the difficult things to light is human skin. Most other objects are easier because little variation exists in their surfaces. But with skin, you have a hundred different shades, hues, and textures.

Most artists paint on canvas; we had to paint on skin. So how do you make skin look realistic, natural, and, in this case, better than it looks in real life? The answer: specialized lighting.

An ad agency was asked to do several commercials for a restaurant. Nothing was really odd about this; restaurants make commercials all the time. This client had a great logo of a bull snorting steam from its nostrils. They wanted to use the logo in the commercial, but this shoot was going to be different. The logo was to be spray-painted on a young woman's belly. As she moved, the bull would also. What could I say? The client is always right.

A beautiful model named Joy was hired to be our human billboard. The first task was to create this logo on her body. The bull logo had been cut out as a stencil on UAL label adhesive paper.

(Continued)

This laser printer label paper is used to attach information onto videotapes. We purchased a few sheets of this adhesive paper (without the cutouts for labels) and drew the logo on the surface with a pencil. Carefully, with an X-acto knife, the portions of the logo that were to be painted were cut out. The portions that remained skin colored would be left alone. This way the bull had depth and shading: flesh and red. Figure 23-1 and 23-2 show how the process was done.

This adhesive stencil was then peeled from its backing and attached to our model's stomach. Pieces of cardboard were used to protect her from paint spray. The director (he had the steadiest hand) applied the red spray paint in short bursts. The aerosol paint was a skin-friendly acrylic facial paint and had very little odor. The color was rosewood, the exact shade of a car I once owned. Luckily, this paint adhered better to Joy than to the fenders of my car.

After a few minutes of drying time, the adhesive paper was *slowly* removed. Much like bandage removal, the art is in the speed, or lack of it. If the paper is removed too quickly, pieces of paint fly off with the adhesive. If too slowly, the adhesive pulls the skin and is uncomfortable. Explain this process before you actually do it to your model. It's

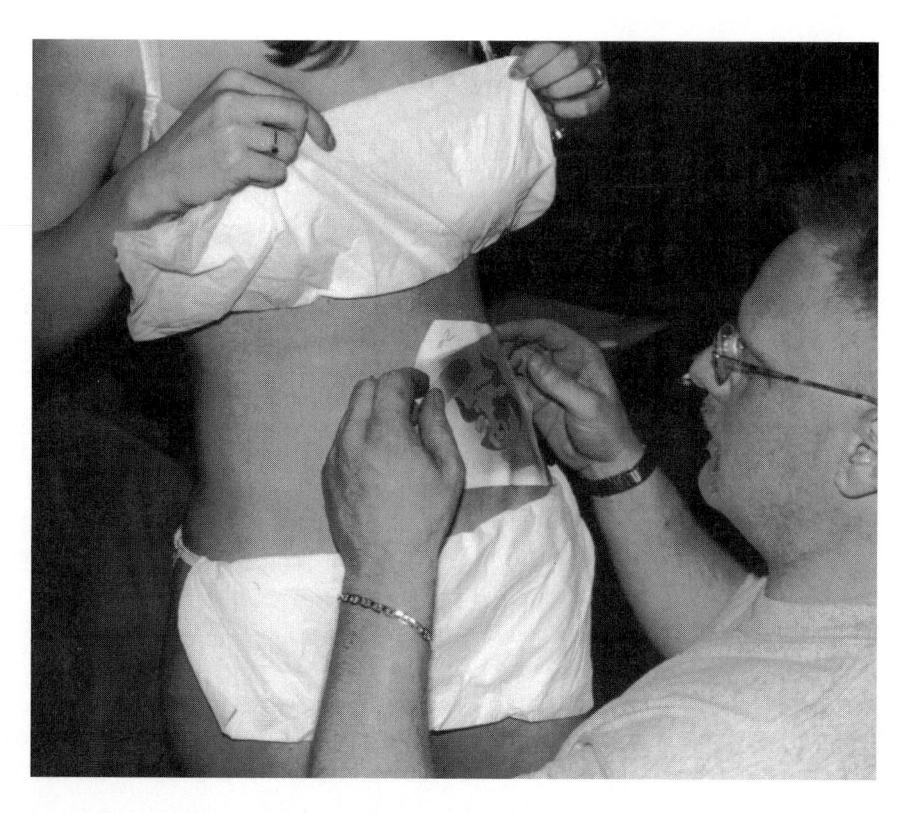

Figure 23-1

Stencil applied to skin

Figure 23-2

The completed tattoo

better for her to know before you just yank and she has no clue what to expect.

Once removed, the red logo looked perfect, much like a fresh, temporary tattoo (because it was). However, because of the adhesive paper, the model's skin was also red and irritated behind the logo. Extremely meticulously, the makeup person had to apply Cover Girl liquid makeup and blend it in behind the bull logo with a narrow brush. The stencil worked perfectly; no spray paint was anywhere it shouldn't have been. The only thing we hadn't considered was how Joy's skin was going to react to the adhesive paper. She was not a plastic VHS shell, but at least no sticky residue remained.

These tattoos can be applied almost anywhere and wash off with soap and water. If a little time is spent in the preparation, you will have an image that looks like the permanent kind (just don't use car paint like I first did; that's semipermanent).

#187 How to Light for Commercials

The wise people in the crowd would say if you want to light something easily, just go outside. But in this section I'll talk about lighting inside and outside.

As you probably know, lighting helps create the mood. Although few commercials are dark and foreboding, if that's the look you want, light it that way. Most, however, are lit with high key lighting. Everyone's happy and well exposed; nothing's dark and hiding off in a corner somewhere. You know where every shadow should and will fall. This is one of the only areas in commercial production where you will have total control. See Figures 23-3 and 23-4.

One of the least used and most important lights is the kicker or backlight. This light is used to separate the product from the background. This diffused light should just hit the upper tip of the product so it is modeled in three dimensions. Without a backlight, the product blends in the background and looks flat and dull. But with a slight highlight, the product will have depth.

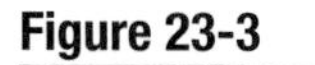

Figure 23-3

A product lit by using rim light or back light

Figure 23-4

Another product that looks appealing because of lighting

#188 How to Tell a Story Effectively in 30 Seconds

Remember the old joke: What's the longest word in the world? The answer: the one that comes after "and now a word from our sponsor." A commercial is really a very short space of time to tell a story. But if you only have 30 seconds, you must make every moment count.

The biggest drawback about a 30-second commercial is that it only lasts half a minute. That's not really a lot of time to tell much of a story. Although when the public views the completed spot on the air, they never realize that you told them the when, where, how much, and why in less than a minute.

Several things go through my mind when I'm creating an idea for a commercial. The first obvious choice is to find out what the product is going to be. As an example, let's say that we have 30 seconds to try to sell a bar of Luigi's Soap.

When I write copy for spots, I try to use the product name as much as possible without it sounding silly. "I use Luigi's Soap because Luigi's Soap is the cleanest soap that a Luigi's Soap Company sells." That gets the point across, but it's rather redundant. Instead I would say, "For the cleanest soap around, use Luigi's." It's more to the point and the product is still mentioned.

Now that we have a product, we have to come up with some kind of story to let the world know about Luigi's Soap. In college, you are always taught about the two different ways to sell products: the hard sell and the soft sell. As the name implies, in a hard sell spot you are usually trying to cram the product down someone's throat (not a good thing to do with soap). The voiceover narrator or salesperson is usually screaming and yelling for you to buy the product. In soft sell, a gentler, relaxed approach is taken. You are still selling the product, but you are doing it in a sneakier but not underhanded way. I'm sure a thousand examples of each type of "sell" come into your mind.

In my mind, humor usually sells best. People will remember a commercial a lot longer if it's funny. Along the same lines, people will also remember a spot if something out of the ordinary happens. How many car commercials do you remember when you just see various models driving by? If one of those same cars morphs into something else, climbs up the vertical side of a building, or becomes dust and vanishes, that sticks in your mind.

So how do we get Luigi's Soap to stick in the viewers' minds? An easy way to write a script is to watch old reruns of *Bewitched*. Darrin always has to come up with an idea to please Larry, save his job, please the client who wants to leave, and keep Endora from turning him into an artichoke.

Once I have a great idea, I'll begin to develop the script. In this case, I'll use a soft-sell, humorous approach. Because our budget is less than one million, we will also try to sell the soap cheaply to the public.

The next big decision is whether to use voiceover narration or show someone speaking. Although narration is the easiest, least-expensive approach, I want to tell a story on how people use Luigi's Soap and show just how many uses this marvelous soap has.

The first place we start is in the supermarket where we see a bar of Luigi's on the rack with the other soaps. We hear a female voice, off camera, say, "Grab a bar of soap, honey." A male hand pulls the soap from the shelf. Okay, now we have the bar of soap.

Next, we'll dissolve to a mother washing her baby in a basin. The baby is almost covered with bubbles. Conveniently, a wrapper of Luigi's Soap is lying beside the basin. (When you open a new bar of soap, don't you leave the wrapper around on display for everyone to see?) The mother is singing to her child as he is being bathed. We dissolve again to a young boy taking a bath. He had been out playing and is dirty. His mother yells " . . . and don't forget to use the Luigi's Soap!" (Notice how I snuck that product name in? I didn't slam it down your throat!)

We dissolve again to a shot of a chimney sweep. The woman tells him, "We've never had the chimney cleaned before; it's got to be really filthy in there." The sweep says, "Don't worry; I've got special stuff just for places like this." The woman walks away and the sweep pulls a bar of Luigi's out of his sack. (In the visuals, he can look around and then kiss the bar of soap. I'll use whatever adds to the humor.)

We dissolve again to a car wash. A filthy black car pulls up, and the girl inside the car fills the coin-operated machine with quarters. We cut to a shot of a workman opening a bar of Luigi's Soap and dropping it into the soap dispenser slot in the car-washing machine. When the woman comes out of the other end of the car wash, her car is white.

Slowly each scenario is getting a little less believable and hopefully more humorous. I want the viewer (and the client) to realize that this soap can be used to clean anything! The dialogue or voices you hear in the background need to accentuate the visuals. If the visuals don't need someone speaking (like the car wash scene), don't use them. Sometimes the visuals need to stand by themselves. And if they didn't bathe, they'll always be standing by themselves.

#189 Review: The 10 Most Important Things to Remember When Writing a TV Commercial

1. Use the product name in the spot as much as possible. The client will like you more when you do this.
2. Determine if your spot is going to be a hard sell or soft sell. Once that is decided, follow through; it isn't nice to switch back and forth between sell types.
3. Try adding a touch of humor in your writing. Notice I said "a touch." No silly funeral home commercials. The viewer will remember a humorous spot longer than a serious spot.
4. Determine what type of audio you want to use: voiceover, someone on camera speaking, or both.

5. Have the humor or pace of editing build throughout the spot. Let the end have a lasting impression (especially when writing a cement commercial).
6. If the visuals can stand by themselves (can you say it without words?), don't use words to bog you down.
7. Write more scenes or shots than you will have time for in the length of the spot. Sometimes things look great on paper, but don't work out when shot.
8. Try to mention only the good (positive) things about the product. If it's expensive, don't mention its price. Mention how much more the competition's product costs.
9. Only tell the viewer what you need to tell them; don't tell them too much. Let them be intrigued enough to want to see more or rush out to buy it.
10. Try to come up with as many ways as possible of saying the same thing. The client is bound to like one of them.

#190 How to Shoot a Commercial

Hopefully, by the time you're ready to shoot, you've already solved the pre-production problems up to this point. You had a great idea, wrote a great script, developed storyboards that would make Spielberg proud, hired the talent, scouted all the locations, and hired the best crew in the business. Now you can shoot the spot.

As you've heard a thousand times in school, the camera is the eye that lets the world see what you want them to see. Because we still haven't completely finished that Luigi's Soap spot, let's use that as our shooting example.

As you remember from your childhood, the first shot is when the male hand grabs the bar of soap from the shelf. When you are first introducing the product, it's better not to shoot it in some bizarre angle, unless you want to get the viewer sick early on. If the product is shot in a medium close-up, at the viewer's eye level, and not framed in a strange manner, the world gets its first glimpse of Luigi's Soap in a pleasing light.

In order to draw the viewer's eye to the right spot in the frame, I can zoom into the product slowly, have other less colorful products around mine, have my product better lit than the others, and also have 5,000 more bars than the competition. Although I will show the competition's product in the first shot, I'm not really doing it in a flattering light. Just like magpies, the viewer is attracted to bright, shiny, colorful, and bigger items in the frame. Instead of just choosing one of these methods, I'm going to employ all of them.

As the male hand reaches for the soap, the camera slowly zooms into the product. When I say zoom, I don't mean the camcorder type that causes whiplash. An extremely slow zoom can be a very effective tool. All the other soaps around Luigi's are less colorful and the lighting is shaded (flagged) over the competition. Besides, nobody wants to buy a dark bar of soap.

For the next scene, we rent a baby and mother, and set the soap wrapper in the shot. The camera will slowly dolly past the mother and child. This slowly moving, right-to-left movement adds fluidity to the shot. When you are showing something as intimate as a mother bathing her child, you don't want to race through the shot. I want love to be oozing out of this scene. A slowly moving dolly adds romance, believability, and gentleness to this shot. And to make things even better, we get to see the soap wrapper in the shot as well.

In shot number three, we also have a bathing scene. In order to keep the censors happy, having a slowly moving dolly wouldn't be as effective in this shot. This dirty boy doesn't need the romance and intimateness that a mother and baby need. The boy would probably listen to rock music because he's tough, dirty, and all-boy. In this case, a static shot will be fine. To make the boy look more imposing than he really is, I will shoot it slightly lower than his eye line. I want this 9-foot-tall 10 year old to have the respect he needs in the frame. That's why, when his mother yells for him to use the soap, his look and the camera look says, "I'm too big to be told to do that, Mom!" All these angles and moves need to be subtle. We want the viewer to recognize them without calling attention to them.

When the chimney sweep does his scene, the camera is at his eye level. We really don't need any moving camera. Through editing, the effectiveness of this scene will be created. When the sweep and woman speak, they are shot in medium close-up. When the characters are first introduced, a two-shot of both people is necessary to establish the characters in their surrounding. Once they are introduced, we can identify more with them through close-ups. You will never identify with characters in a commercial if you see them in a long shot.

To be intimate and create this identification, we need to get close to them. This happens the same way in reality. You won't get close to someone across the room; you need to *be* close to them (don't forget to brush your teeth). Just consider the camera as an extension of yourself. How would you react in a situation? What do you want to see happen? You'd be surprised just how much common sense comes into play when shooting.

The car wash scene is more complicated. We have quite a bit of information to get across without any dialog or voiceover. What we show the viewer will make or break the shot. Remember, begin with a long shot to establish the surrounding. "Oh, look at that dirty black car and that shiny new car wash." Cut to a close-up of the woman as she reaches to put the money in the slots. "Gee, she's attractive. How did she let her car get that dirty? She's going to need sand blasting to get that thing clean!" All these things can go through the viewer's mind because we are shooting her in close-up. We really don't need a close-up of the money going into the machine. From our first angle of her, this should be evident. But to make my life easier, I will shoot a close-up of the machine. I don't have to use it in editing, but if I decide to, I'll have it.

However, we *do* need to show a shot of the man inserting the soap into the soap dispenser. If we cut this scene out because of lack of time, the humor of the joke is lost. Once again, we establish the soap man in his environment with a medium long shot. But when he puts the Luigi's Soap into the soap dispenser, we need to *see* that. My choice and yours . . . a close-up.

#191 Review: The 10 Most Important Things to Remember When Shooting a TV Commercial

1. Save the experimental shots for later in the spot. The first shots should be standard enough to let the viewer know what's going on.
2. Try to establish your shots in this order: an establishing or long shot, a medium shot, and then a close-up. This isn't gospel, but it's what the viewer is used to seeing.
3. Keep zooming down to a minimum. You don't need to add whiplash in a spot. If zooming is necessary, make it slow.
4. Keep the things that should get the most attention very large and dominant in the frame. The eyeball never lies!
5. The height of the tripod is very important. If a nine-foot-tall actor is looking in your eyes, shoot it that way. Low- and high-angle shoots are effective, so use them accordingly.
6. Use a tripod if you want a steady shot; use a hand-held camera (or a Steadicam) if you want that type of effect. Don't be lazy.
7. Use a moving camera whenever possible. A slow dolly, jib, pan, tilt, or zoom can go a long way to adding emotion to the spot.
8. The camera is the eye of the viewer; only show them what you want them to see. Make them look only where you want them to look.

9. Although you don't really have to follow tradition and do what people expect, if you deviate from the norm, be ready to explain why.
10. Don't arrive on the shoot without knowing your camera angles. Things can be revised on the set, but it's no place to first start thinking about camera angles.

#192 How to Edit Television Commercials

Some people think that editing for a television commercial is more difficult than editing for some other type of project. Each project has its own set of difficulties, and a commercial is no different. I'd rather not say that the editing is more difficult; it's just more precise.

When watching a feature film, the director can linger on a close-up for a long time if the story or action demands it. To really know that a lead character is suffering, the camera slowly pulls into the pained actor's face. We can see the worried expression on his face, the tear-stained eyes, the slightly quivering lip. If it takes four minutes to get that point across to the viewer, so be it. We don't have that luxury in commercials. In 30 seconds, we've got to introduce the character, product, price, and everything else the client wants to tell. If we can spend more than three seconds on a tear-stained face on a $60,000 car, we're fortunate.

One of the most important factors in editing commercials is pacing. Once it's established, don't change it unless you have very good reason to do just that. If you're editing to the beat of the music and the drum strikes every 24 frames, don't leave a shot on the screen for one and a half seconds. It will stand out like a pig at a swimming pool. Whatever shot is on the screen the longest will stand out from the rest of the pack and leave an impression in the viewer's mind. Imagine a metronome ticking away to set your pace. Be rhythmic, tap your foot, bang your head against the wall, or snap your fingers to the beat. Do whatever it takes to keep it structured.

The hardest thing to do in a 30-second spot is to keep the spot at precisely 30 seconds. One extra frame, just for effect, won't do. This usually means that you have to cut the shots as tightly as possible. If the point gets across in 2 seconds and 23 frames, don't let the shot linger on the screen for 5 seconds. You aren't afforded that luxury. Later on in the spot you'll definitely be able to use that extra time. I've rarely had too much time at the end of a spot.

This has to be one of the toughest things to get across to my students. They'll turn in 33- or 24.3-second spots. I tell them if they learn anything from this class on making commercials, your time limit is 29.5. That is the actual length of a 30-second spot, no more or less. If you send a spot with

#192

less time to air on a station, they will pull it because there will be blank air-time, a taboo in television. If the spot is too long, even by 6 frames, the system will stop it at 29.5 and no one will see the remaining visuals; it's that precise. This was the toughest thing I had to learn in TV, and it needs to be mastered at the beginning.

From the very beginning of the spot, keep track of its length. After you lay in that fourth shot, how much time do you have remaining? Once again, be very precise. If you only have 20 frames left at the end, that still can be an effective amount of time. If the great closing shot lasts 28 frames, then find someplace where you can cut 8 extra frames off a shot. The last thing the viewer sees will probably remain in his or her mind the longest.

If a great shot lasts 26 frames (we're talking about a very short shot), and you have only 18 frames to fill, try changing the speed of the shot. An image that's running at 103 percent speed won't have a noticeable onscreen speed change if it only lasts 18 frames. The same thing applies if you have to make a shot last longer than the amount of footage you have. Although you will rarely have too much time, it can still be a problem. A shot running at 96 percent is slowed so slightly, only small dogs and bright children will notice.

The opens and closes of commercials are extremely important because they are the first and last things people see. This may sound like common sense, but many editors overlook this. Should you fade into the first shot or cut into it? Although I don't want to play director on your editing session, a shot that's faded in is much softer and smoother, but it will eat up some screen time. Try a 15-frame fade instead of 25 frames. You've just gained 10 frames (they can be redeemed for valuable prizes later on). The same thing applies to the end of the spot. I prefer not to fade out at the end since I don't usually have the time because I've crammed too much into my spot already. If your spot is running back to back with another spot, chances are that other spot will fade in immediately after yours cuts to black. Through a trick of the eye, the viewer will not see this as two different events. They will think that one went smoothly out and the other came smoothly in.

In this era of electronic editing and linear versus *nonlinear editing* (NLE) systems, we have thousands of different effects or transitions from which to choose. While working on a low-end nonlinear system, I counted 322 different wipe effects that were installed in the system. I'm just dying to use the one where the tuna fish eats the boat and the fisherman. I haven't even discussed all the dissolves, page turns, defocusing, mosaics, posterizations, slides, and four million other DVE moves that exist. All these effects can be tempting to an editor. Just remember, if it doesn't help advance the

#193

story of the spot, don't use it just for the sake of using it. Using it might make your production look less professional even if the effect looks really groovy.

The other factor to remember again is time. A 45-frame dissolve is much slower and more time consuming than an equally effective 20-frame dissolve. If your story needs a dissolve or effect of a certain length, use it, but just do it sparingly or remember the price of time you may have to pay.

When a new effect comes out, everyone and his brother uses it. Remember the morphs? (No, they're no relation to the Smurfs.) At first, they appeared new and exciting. Then you rarely saw a spot that didn't have them; now you won't see them except in a science fiction movie.

If you want to make a statement by cutting a spot a certain way, do it. But just remember, if you are doing something new and trend setting, you'll be questioned about it forever. Make sure you have a good reason for using it. Saying "the devil made me do it" no longer works (unless you're Flip Wilson and he passed away).

I don't have the space for getting into narration versus lip sync or whether to tell a story with great acting or just show the product in its environment, but I do want to talk about onscreen text. Character-generated text is just that. It's generated to enhance the spot. Should you use all uppercase characters or mix the two, right or left justified, center screen, upper third or lower third? All these questions depend on the look you and the client want. No one way is better than another. Right now, moving text (very slowly expanding, contracting, or sliding across the screen) is extremely popular. Like morphing, if you have text on the screen, it must be moving. This trend will also change eventually. Just remember that the viewer has to be able to read the text, so don't cram the screen with too much written information or have it moving in too many directions at once or their eyes will bleed.

If text does move, the viewer's eyes will follow it. The human eye will follow movement first; then it will scan the rest of the frame. Most importantly, check and double-check your spelling. I've learned that the hard way. A decimal point in the wrong spot can also be disastrous. Being the editor on a commercial production, you have the power to make or break the spot. This is what separates award-winning spots from the boring ones. I've seen dull spots about dull products jump to life because of great editing.

Because of your limited amount of time in a commercial, great shots that were recorded in the field may end up on the electronic cutting room floor. You have to pick the best shots and use them in the shortest amount of time possible while still making it interesting to the client and home viewer. Like I said at the beginning, editing commercials is easy!

#193 Review: The 10 Most Important Things to Remember About Editing a TV Commercial

1. Try to establish a rhythmic pacing and stick to it for the duration of the spot. Sing the song, "I Got Rhythm."
2. Try cutting the spot to the beat of the music. In no time, people will be dancing to your soundtrack, and nonlinear systems make it extremely easy to edit.
3. Edit as tightly as possible. Keep each shot on the screen only as long as necessary to get the point across. If you extend past your 29.5 seconds, you probably didn't edit tightly enough.
4. Don't be a sloppy editor. Watch for mismatched frames, dropouts, high video and audio levels, and so on.
5. Keep close track of the running time of your spot. If you don't, you may run out of time before you've finished.
6. Change the speed (running time) of a shot slightly to extend it or shorten it.
7. To fade or not to fade, in or out, that is the question. How much time do you have? What effect do you want?
8. Use bizarre and unusual electronic effects sparingly. Old-fashioned dissolves or cuts can be just as effective.
9. Don't be afraid to do something out of the ordinary. You may have discovered a better way of doing something. Just know the price of fame and fortune you must pay.
10. Use text sparingly. This is television; a lot of people don't like to read. But if you use text, make it look interesting (but not like subtitles).

#194 How to Record Sound for Commercials

When I ask people who will be on camera where they want the sound person to stick the microphone, they usually tell me. How do you get that perfect sound effect, voice, or sound, and where do you put the mike to get that sound?

Let's start with recording people, because they are the most difficult. Since most people are living, breathing, and moving examples, they can be difficult to mike. If you use a shotgun mike on the camera, the sound can be hollow, and if the room is too big, the sound will echo. On the bright side, a shotgun mike has a very limited pickup pattern; distracting noises in the background can often be missed with this type of mike. In a live, noisy environment, a shotgun mike won't pick up the talent without getting other noises behind them. If the person is talking and all the distracting noises are behind the camera, then this type of microphone may work out.

Another plus is that this type of microphone isn't attached to anyone. You won't hear clothing rustle, heavy breathing, or muffled sound. A shotgun mike is out in the open so everyone can see it. Obviously, you'll get a different quality of sound from different shotgun mikes, so experimentation is best. Cheap mikes won't sound as good as more expensive models, and you will get what you pay for.

The lavaliere mike is usually pinned, attached, or stapled on the talent. Although much smaller than the shotgun, a lavaliere shouldn't be underestimated. Unless you want people to see the type of mike you use, a lavaliere should be hidden under clothing. This creates the problems I just mentioned. Cotton can be very loud and obnoxious when placed under or over a microphone. As the tiny molecules of thread scrape across the delicate surface of the microphone, sound people have compared this to fingernails on a blackboard. On the plus side, this mike is usually much closer to the talent and makes a better pickup device (don't try it in bars).

When the mike is attached to the talent, they immediately become more self-conscious that they have a mike on. This sometimes can take away from their performance. My rule of thumb is if you're recording close-ups, mike close up. If you're shooting something in a long shot, a shot gun will work or use a lavaliere and make sure the level is turned down. Once again, use common sense when recording sound. A whisper shouldn't be recorded at full volume, and a shout should show up on the VU meters.

When recording sound effects from inanimate objects, a directional shotgun mike is best. This type of recording situation allows you to just to point and record. Record the levels as hot as you can without distorting. You can always lower the sound level of the effect in editing. Get as much of the sound as possible, and record more than you'll ever possibly use. If you're recording sound in Antarctica, it's expensive to return just because you got too little sound. Don't record in the field with any filters or equalization. These types of additions (or subtractions) to the sound are best left for postproduction. If you record an effect with a filter or equalization and decide not to use it at a later date, you'll be in deep trouble.

Play with where you want the sound to come out (which speaker, left or right). If you can have an effect move from one speaker to another, do it. A little audio sweetening can make a production come alive. And speaking of sweetening, be careful on the Foley effects. A lot of spots overuse and overmodulate a created effect. Use common sense; if an effect should be amplified or recreated, do so at the appropriate level so that it will be noticed. An effect that overly calls attention to itself is an effect that didn't work.

Room tone, or ambience, also shouldn't be neglected. Your job as a soundperson is to give the most complete production sound available. That means nonmuddy vocal tracks, location sound effects, and ambience. When the shot is completed, ask for at least 30 seconds of quiet as you record the room tone with everyone still in their positions. I usually speak with the camera operator early on and ask if, at the end of the shot, he or she will call for quiet. If the camera operator does this, even nonprofessionals will think he or she is shooting and will remain quiet.

#195 How to Survive a Log Flume Commercial

When one conjures up an image of an amusement park, several things usually come into mind: hot, humid, blistering heat, screaming crowds of children, the smell of greasy hot dogs and cotton candy, waiting in long lines to get on a ride, and, finally, lots of fun for the whole family.

I was asked to shoot a short video on the newest ride at an amusement park. This ride has to be one of the most exciting and visually appealing rides in the park. It consisted of two four-foot-wide enclosed plastic pipes that snaked around a wooden enclosure. Climbing to 75 feet, the passenger would descend at an alarming rate and be spewed out into a trough of water. Pipe number one was blue, like the color of a plastic tarp. Pipe number two was jet black, totally opaque. A small stream of water ran down the inside of the tubes for lubrication and momentum, and the passengers would ride in a two-person rubber raft. With operators at the top of the ride unloading empty rafts off a conveyor belt that allowed them to dry out, they would methodically help patrons into their rubber contraptions. At the end of the ride, additional operators would assist in removing the dazed occupants from their wet vehicles and reload the conveyor belt. It appeared to be a vicious cycle.

I came up with the concept of shooting the entire ride subjectively and then intercutting various aspects of the ride, such as people getting in rafts, riding the curl, and so on. Unfortunately, the sky on the day of the shoot was very overcast and ominous looking. The grayness of sky didn't really lend itself to the happy smiling faces of children and water. In addition, because it was late spring and the park was just open on weekends, the temperature was cold. There's nothing like riding a water ride when you can see your breath. It was also difficult to give the shoot the warm, summer look when all the guests in the park were wearing sweatshirts and long pants. Only the brave souls were wearing shorts and T-shirts.

The client then had the usual suspects (all wearing shorts, strangely enough) sign release forms. We taped the wide exterior shot of the ride at an extreme low angle to make it look very dominant in the frame. However, the gray sky with evil-looking clouds still made the ride look scary.

We decided to use a Tiffen Graduated filter to change the color of the gray sky to a more appealing hue. With the orange-brown color at the top of the filter and clear glass on the bottom, the clouds, although still dark, had a warmer, friendlier look. With the sky looking more pleasing, we had our long exterior shot.

We watched as people exited the tube and noted their head clearance. We also timed the length of the ride to determine when people would be spitted out of the tube. We then climbed over to the open trough where the riders exited and up onto the mouth of the blue tube. Holding our Betacam camera over the opening of the tube, we leaned on the protruding metal lip and watched the video monitor. The diameter of the open tube was four feet from side to side and four and a half feet from top to bottom. We wanted to make sure no one would hit the camera as they exited the tube at 90 miles per hour. Talk about a bug splattering on your windshield.

As the human shadows raced down the tube, we could tell the exact placement of each rider. It was exciting watching the black splotches slide around the tube as their muffled screams announced their positions. The camera would be lowered into position and we would have 11 seconds until they exited the tube. The mike on the camera captured their screams as they saw the camera protruding from the mouth of the tube. If the camera were any lower, it would be the last thing that the rider would see before they died.

We recorded at least 10 different people exiting the mouth of the tube. Our wrist strength was tested as we held the camera over the opening. A trough about 50 feet long was connected to the mouth of the tube. As the riders exited, they would slide down the trough and hit some vinyl that would slow them down. A 15-foot section of the trough was all we had to work with. Once the rider hit the vinyl, the ride was essentially over and made an uninteresting shot.

The camera operator and I stood on the left of the trough. We put the camera down directly on the vinyl. As the rider came barreling out the mouth of the tube, I wanted to capture that split-second instant when the rafter came directly into the camera's lens. Much like trying to videotape a freight train careening toward the camera, I underestimated the speed in which these raft riders of death were traveling.

While looking at the monitor, we found a perfect spot for the camera. As soon as the rider came within 10 feet of the camera, the operator and I would hoist the camera into the air to avoid it being crushed. I believe this is where my brain stopped working momentarily.

As the camera remained right on the ground in the middle of the trough, tethered with a line to the monitor, the camera operator held the camera by its carrying handle. I had my arms around his waist to pull him to safety if need be.

Now came the moment of truth. We scanned the tube for shadows. We heard the ride operator at the top announce that someone else was on the way down. From the time of her yell, we had 26 seconds until impact.

I caught my first glimpse of the rider's shadow halfway down the tube. Time was slowing down. The camera operator was holding the camera and waiting for my signal; he couldn't see what was happening. As the clock ticked, I thought of something I should have realized earlier. How are we going to get a clear shot of these people and still pull the camera away in time? I really should have thought this out better before we set up. As I gave the countdown before impact, I told the camera operator to pull the camera when I said one! By the time I hit number three, the riders had already exited the tube. That couldn't be! After timing 8 different people, it always takes 11 seconds for them to exit the tube.

Then I realized why these riders had exited earlier. A hefty gentleman was in the back of the raft and his 200-pound son was in the front. Mass times volume times fat people equals . . . I had to think fast. Before I got to say two, I told the camera operator to pull the camera away as I grabbed and yanked him by the waist. As gravity started to take hold, the last thing I remember seeing is the look of horror on the father's and son's faces. Here they were enjoying the ride, and the last thing that they expected to see when they exited the tube was two people with a 26-pound camera directly in their path.

As we arched into the air, we avoided being hit by less than five inches. Although no obscenities were uttered, if we had more than a split second to think, I believe we could have thought of a few.

The rest of the shots of the ride had to be easier. All we had to do now was get some shots of people boarding the rafts at the top of the ride. We grabbed our gear and began to climb the wooden stairs that led to the beginning of the ride.

After we climbed the 461 stairs that led to the entrance of the ride, we set up for the boarding shots. Of course, having a camera and tripod in an area full of kids led to a lot of stares (even more than the number of wooden ones we climbed).

The client then asked if we wanted to get some footage of the black tube that was loading 20 feet from our present position. The black tube offered the rider a pitch-black ride, tossing and turning, until they exited into daylight. We could show people boarding the ride, but it would be difficult to achieve a subjective shot in total darkness.

The camera operator climbed on top of the blue tube and got a bird's-eye-view shot of riders entering the tube. Hand holding the camera, he would swing down into the mouth of the pipe as each rider's journey began.

After lots of reaction shots and loading sequences, we were ready for our own subjective shots.

The camera operator and I piled into our rubber raft. A small puddle of muddy water lay in the bottom of each raft. The cold water wasn't as comfortable as I had thought it would be. It felt more like a wet diaper. The camera was positioned on the operator's shoulder and I sat in the rumble seat with the ever-present monitor. Within seconds, I was pushed and we were flying down the tube, banking and sloshing around. I thought that we were going to burn out in reentry when we attempted the sideways loop the loop. Because of gravity or the G-forces we were experiencing, it was impossible for the camera operator to keep the camera straight or level. In addition, it's difficult not to scream when moving through an enclosed tube at breakneck speed. We screamed just like the rest of them. The water soon had us wet from head to toe.

Back up the 461 stairs and into the raft again, but this time we used a smaller, digital camcorder. This camcorder weighs less than two pounds, including the battery, and is much easier to hold while sliding down a wet pipe.

From the wide shot of the exterior of the ride to the final shot of the tube belching out its lunch, the viewer experiences 11 minutes of pure adrenaline-filled fun. Now if only we could do the ride without the cold, muddy water in the bottom of the rafts . . .

Chapter 24

Human Challenges

#196 How to Deal with Humans

Although not the most difficult thing to deal with, humans do pose problems that animals and inanimate objects don't have: higher intelligence. Because we humans can think and reason, we will cause more problems for the director.

This really is a problem mostly because people think too much. You must relate to nonprofessionals, attractive people who may think they are a better example for representing the human race, on-camera people who were bred just for that purpose, prima donnas who were put on this planet to annoy their fellow man, and people you have to track down that you cannot pay for their time.

The last item covered in this chapter dealing with human challenges is working with multiple crews within the same project. The viewer prefers some type of routine within the video, such as shots in focus, people being framed within the shot, and some type of continuity from shot to shot. Once these challenges are met, you will have no other problems, at least in this chapter.

#197 How to Direct Nonprofessional Actors

One of my favorite parts of producing or directing is being able to cast the talent. Like Cecil B. DeMille, you look at thousands of faces and biographies, and pick that ones that interest you. When you work with professionals, you can find the exact type of character you're looking for; that's what professional actors do—they act. However, sometimes you aren't offered the luxury of being able to hire outside talent.

The on-camera people are what the viewer sees; they never see all the work that goes on behind the camera. If the client's

Uncle Vinny does a poor job of being a convincing customer in the video, then the public is going to have a hard time believing him as an actor. The client then comes back to you and says the video didn't work and was a waste of time. If they had let you do your job in the first place, the program probably would have been well received.

I'll try to take this one step at a time and cover some of the possibilities you might encounter in a situation like this. The first step is to get your auditionees together and explain what you're trying to achieve in the video. I like to role-play all the parts to everyone so they know what I'd like to see them do. You should have a clear vision of what each character must do, say, and act like. It's almost impossible to get a spectacular performance out of someone if you really don't know what you want. Once you've done your routine in front of the cast, don't ask if anyone has any questions at this time. Everyone will have their own expert option on how they could perform the part better and add more life to the video. Instead, I prefer to get each actor aside individually and go over their part one on one with them. I do it as a group effort first so everyone can see how each part fits together as a whole, but the one on one is where the little nuances come to life and when I ask if the actor has any questions.

With this one-on-one approach, I act out what I see the character doing, and then I ask the actor to try his or her hand at it. I find it better if I perform the role first. They usually have an easier time mimicking my performance than doing their own (often wrong) interpretation. A role can be played in a million ways, but if you have something specific in mind, save yourself a lot of time and show it to them.

If any of the characters have speaking parts, that involves a little more skill and patience on your part. Actors who are used as extras aren't much of a problem, but if they have to speak, they require more instruction. Have them say their lines to you in a conversational tone, just as if they were talking to you. Some will try to project their voices and talk very loudly, while others will overact as if they are onstage in front of thousands. I tell them that television is a very intimate medium. It's very conversational. If two people are sitting at a table talking quietly, that's just how they should do that. The only difference is that a camera will be close by. The three most important things for nonprofessional actors to remember are to be yourself, be yourself, be yourself.

Now that they have mastered the dialogue, don't try to have them do it all at once. If you break it down into smaller bits, they'll have a better chance of doing it right and feeling more comfortable with it.

Once every actor is familiar with what he or she is supposed to do, shoot everything in sequence. It's very difficult for a nonactor to shoot something from the middle first, then do the end, and then the beginning. You'll get more out of his or her performances if you shoot chronologically.

As a director, the talent needs to know you're in charge. They can't say "cut" in the middle of a take because they don't like it. There will be laughing jags and nervous sessions when they just have to get it out of their system before you get anything done. Don't treat them like children either (unless they are children), and try not to talk down to them. They are excited about doing this and they will remember the experience long after you've forgotten it. Try to make it an experience they will enjoy. Let them have fun with the role and you will get much more out of them.

Always rehearse several times before you roll any tape. The red light on the camera, the quietness of the set, and them being "on" for the first time can make them uncomfortable. After you get a realistic performance out of them, roll the tape without telling them that the camera is on. If it's a great performance, I'll tell them to do it "for real" this time. If it doesn't work out because they know the camera is on, at least you got a good take earlier. If you get even a better performance out of them, then you have a safety.

Always expect the unexpected and don't blow up at someone for not getting it right after 60 takes. They're very nervous, they want to please you, but some people don't have it in them. Instead, take them aside and try altering the script slightly. If you embarrass them in front of everyone, you've lost them as an actor.

#198 How to Choose the Best Model Without Being a Bad Example

When shooting a video, it's difficult to tell a lot about a model's talents by looking at a still-photo headshot. To make my life easier, a local modeling agency decided to have an open house.

All area TV and film producers, ad agencies, and bridal boutiques were invited to a two-hour open-house, meet-the-models session. In my line of work, being able to meet and talk to the models is important; many times print advertisers can just look at a photo to see if the model will, in fact, work out. Most of my talent has to have acting ability.

The model show began with 45 models (men, women, and kids) walking down a runway, telling a little about themselves, and showing some special talent (singing, dancing, and so on). This was helpful to see if they could actually walk and talk at the same time. Some of the models were great to look at but had a hard time walking without tripping over their feet.

After this 30-minute session, the owner of the agency introduced the attendees. Over 30 people had been invited and promised to attend this event, but only 4 people actually showed up. As the owner introduced us, she pointed us out. Two people were there from a bridal salon. They were only looking for male models to pose in some print ads. Another gentleman, who looked like Steven Spielberg on a bad hair day, was there representing a print agency. He didn't really need anyone at the moment but was just looking (this wasn't a new car dealership), and then the owner turned her attention to me. "And this is Chuck; he's a television producer. He's looking for people to be in TV commercials." If she had said that I was Brad Pitt, had dropped gold nuggets on the floor, or said I was filthy rich and senile, I wouldn't have gotten more attention. She mentioned the names of no one else; she glossed over what they did and what they wanted, but she singled me out. I worked for a local TV station and make local commercials. I wouldn't be doing national, million-dollar-budget spots until next week; all that made no difference.

The owner then continued to say that since the evening was young, if any of the models wanted to talk to any of us, they could. It was at that exact moment that the floodgates opened. All 45 men, women, and children ran to where I was sitting and formed a line. The line extended out the door of the agency and onto the street. I now had my 15 minutes of fame.

While I was watching the show, food and drinks were available. As my luck would have it, I was stuffing a greasy meatball into my mouth with a pink toothpick when the owner announced my occupation. I was interested in meeting *some* of the models, but now I had to meet and talk to every single one. In addition, the way the owner described it, I had immediate openings in several commercials for the right person. Time to impress the producer!

As I tried to swallow my meatball and wash it down with some $2 wine, the first model plopped her portfolio on my lap. Wiping my hands on one of those one-inch-square cocktail napkins, I perused her portfolio. This woman was God's gift to earth. She had the kind of beauty that few have been blessed with, and when you looked up beautiful in the dictionary, her picture was there. And to make matters even more uncomfortable, she was wearing some exotic perfume that would turn a mortal's brains into dust.

The first pictures were the most useful to me. She was wearing a variety of outfits and posing with props. I could see that she was versatile and, with different makeup, could play a range of ages. As I continued to turn the pages, she began to wear less and less clothing in the photos. By the time I got to page 10, she was wearing nothing but a smile.

First of all, it's difficult to look at a naked picture of someone you don't know when she's sitting five inches from you. Second, in local TV, I rarely use naked women (most of our local sponsors aren't ready for that just yet). Although a man doesn't really mind looking at pictures like this (so I've read), they really are of no use to me.

As I began to ask her some questions, my voice squeaked. I'm sure it was because I had swallowed some wine too quickly after seeing her in the buff. Surprisingly, she was very literate. As she explained her aspirations to me, I tried to flip through the remaining 90 nude pictures as fast as I could (I didn't want to appear rude). I then told her what type of model I was looking for in upcoming spots, and she said she could play those roles. If I had said that I wanted a 90-year-old, 600-pound, Asian male with blond hair, she would have told me she could play the part. As I continued to ask her about her character range, if she had any speaking experience, and if she could act, she showed me her range (her acting range that is).

As the next 23 gorgeous women sat down in front of me, showed me their portfolios, and asked me when they would start working, I developed a game plan. I wasn't there to lie to them, promise them parts that didn't exist, or lead them on in any way. Although I was in male heaven, surrounded by the most beautiful women on earth, I was there for a real purpose. I asked them if they would have any difficulty playing a young mother. Most of my commercials center around a mother trying to pick the best product for her family. Would they mind if I made them less attractive so they could play the average young wife, struggling student, or business executive? As I talked with more and more of them, I became more comfortable and less self-conscious that my tongue was going to fall off or that I had a piece of parsley caught in my teeth.

This soon became an excellent learning experience for me. I had the opportunity to interview people en masse, learn a lot more about them than I ever could from a photo, and see what they were really like in person.

The two male models that were in line proved to be more talkative than the women. After crushing my hand with their grip, I got to look at more photos. All the beverages I had consumed in the presence of the women had created a pressing need. Although the bathroom was less than 15 feet away, the crowd of interviewees wouldn't give me the few moments I needed.

In closing, I believe that everyone who uses models in roles that involve acting, speaking, and character playing should meet with as many people as possible. Photo headshots are too dated, people don't really look like that in real life, and you have no way to tell if they can really act. I suggested to each model that he or she get a video resume where prospective producers and directors can see them move, talk, act, and so on. In this video age, we should have that work for us. Even if the tape is done with a camcorder, it will show if the video camera likes them, even though video is far less flattering than 35mm film. I still believe that models should audition in person for their producers/directors, but a VHS tape can help weed out people that would never work out. In my case, people that I had no interest in interviewing from the headshot I saw turned out to be some of my best choices when I met with them in person. If a picture is worth a thousand words, a video should be worth ten thousand and meeting them in person a million.

One last word of advice: Leave yourself a bathroom break. It makes the interview process more enjoyable.

#199 How to Use On-Camera Spokespeople

When someone on camera is speaking to the viewer, he or she is representing your client and should know how to perform the job properly. On a particular car dealership commercial, the client had already hired a spokeswomen for the spot. I assumed she had been screened, auditioned, and at least seen the script—I was wrong on all accounts. She knew she would be doing a commercial, but that was the extent of her knowledge and training.

In order to get a more fluid appearance to the spot, the camera would be dollying around each of the vehicles as the spokeswoman walked and talked. I won't go into the nightmare I experienced laying straight and curved dolly track on an uneven showroom floor; that's a story in itself. With a shoot as moderately complicated at this one, the one bit of advice I can give is rehearse and rehearse again. With a camera moving, a person moving, and the cars hopefully not moving, everything comes down to timing.

After rehearsing the 30-second, 1-shot commercial 12 times, we each had our act in gear. As soon as the camera began rolling, customers came into the showroom out of nowhere. They would see the lights, freeze, and destroy a good take. Why do customers always have to cause problems? That problem was soon remedied by having a salesman watch each door. This also kept them from lingering in the background and staring into the camera.

As the spokesperson walked past each vehicle, she would gesture to the windshield (the client's idea, not mine) and the price would appear. As she gestured, the camera would stop its dolly and pan to the right. However, from takes 23 through 31, each time she gestured she would hit her hand on the sideview mirror. Besides the obvious noise, painful grimace, and blood on the mirror, it just didn't look natural. When someone inexperienced feels comfortable with a gesture, it's difficult to change that. Gesturing to the left didn't make sense, and doing a cartwheel might be too much. We finally came up with a compromise that worked for everyone: a smaller gesture.

On take 41, everything was working perfectly—for a 32-second commercial. We couldn't cut the copy, so she had to say it faster (not walk faster, as she insisted on doing).

By take 67, I was bald, the cameraman had to have his eye surgically removed from the eyepiece, and the client was beginning to think that the whole concept of the commercial wasn't going to work. This is where the fine art of public speaking comes into play (some people call it "kissing up," but I prefer the gentleman's approach). I explained to him that the concept was fine; we just had to make the spokeswoman feel comfortable about doing it—correctly.

After he had been reassured, we continued. Each take through the 70s and into the early 80s was racked with little glitches like sneezes, a coughing fit, a squirrel that somehow got loose in the showroom, and other less interesting problems. After doing more takes than my pocket calculator could handle, we finally achieved the perfect one (the actual number is too big to fit in this book).

Everything worked flawlessly. When the cheering subsided and the hugging was over, we headed back to experience what every director does at the end of a perfect day: going to bed!

#200 How to Work with Prima Donnas

Some people are so full of themselves it makes it very difficult to work with them. Unfortunately, you may be in a situation where you have to do your job no matter how obnoxious the person might be.

I had to videotape a CEO of a huge corporation for a corporate video who loved to be on camera but was having a bad day. Because of his importance, we arrived two hours earlier and had everything set up and ready to go at least an hour before he arrived. Rule number one: Be ready when they arrive.

When he walked into the room, our makeup person asked if she could apply some powder to his shiny, bald head (she didn't use those words, but I'm still mad at him). He promptly pushed her away and said he didn't wear make-up. It didn't matter to me if he came across looking like a ghoul. Rule number two: Ask before applying any type of makeup.

He had reviewed his speech an hour before and began to read through it on the teleprompter. Before he sat on the edge of his desk, we attached a tiny lavaliere mike to his tie and was adjusting his audio level as he read the prompted script. Not liking anything he had approved earlier, he constantly changed his wording so it would flow better. Rule number three: Make sure the script is approved before it is shown on the teleprompter. Some think that once in this domain, the words cannot be changed. Reassure them that you will use any words they feel comfortable saying.

At this point, his anger was building and I should have seen it earlier in order to calm his jangled nerves. Rule number four: If you see someone getting agitated, try to solve the problem as early as possible. You will definitely notice the warning signs: a short temper, fighting with you and the crew, looking at his watch, complaining about the lights, and so on. When he first complained about the script that he approved, I should have asked for a few moments in the room without the crew for him to gather his thoughts. Once the anger builds to a peak, you can't do much to diffuse the situation. Rule number five: Clear the room if you perceive a tantrum about to happen. This is less embarrassing for everyone involved.

As he kept stumbling over his own words take after take, I knew we should have taken a break until he was less frustrated. Rule number six: Don't treat your talent like children, but use the same psychology and give them a timeout (without resorting to those words). When they become overstimulated, they must cool off.

He kept stammering along until he had reached his climax and stormed out, microphone still attached.

Just because a CEO can act like a spoiled child, you and your crew have to be adults in these situations. Do whatever it takes to make the talent happy. We have schedules just like they do, and even though the people in front of the camera may try to call all the shots, don't let them. This is your shoot and you need to remain in control. As mentioned earlier, once you lose control, you will never get it back again.

Take each situation a step at a time and do what feels natural. Bend over backwards so you don't offend them and still treat them with respect no matter what happens. A little bit of patience goes a long way. Never raise your voice even if they are screaming at you. Remain calm and cool, and if they notice that their yelling isn't getting a response from you, they will soon tire and give up. If you lose your cool, the situation has just ended and they won.

#201 How to Get Talent You Don't Have to Pay

For some shoots, you just don't have any extra money to spend, but you still need pleasant-looking people in front of the camera. Where do you find people like this in a world that's full of moneygrubbers?

The first place to look is your own home. Family and friends all know you are in the "business" (not the Mafia), and like Lucy Ricardo asking her husband Ricky, most want to be in your "show." If someone you know or are related to fits the part, give them a chance of a lifetime and let them be in your project. This is an almost unlimited source of free personnel.

But what happens if you are far away from family and friends and need people to be in your video? The easiest thing to do is just ask. This isn't a come-on or a pickup line; you have an actual need (it still sounds sleazy, doesn't it?). Most people, when you first approach them, will not know if you are for real. Always have business cards printed up and hand one to them as soon as you speak.

Phony people can print business cards as well as you can, but it still is the first step. You now have the person's attention. When I need people of the younger persuasion to appear in TV spots, I always go to the local mall. There I find hundreds of willing participants for my experiments. However, when you approach someone in a mall with a business card, that is rarely enough. I need to have a little more clout or I would have been arrested for canvassing.

In that same mall, I wore a golf shirt (don't just use a T-shirt with a saying on it) that had my station's call letters printed on it. Business cards can be printed on anyone's $69 inkjet printer. Golf shirts cost a few more dollars and the illegitimate producers won't spend that kind of money. In the fall and winter, I also wore a jacket with a four-color logo on the back. Yes, I was a walking billboard for the station, but at $200 a pop, I had even more believability.

Of course, times occurred when we were desperate for someone to look longingly at the client's product and needed someone who wasn't at our location. A quick drive down to the mall and like wolves, we were on the prowl. This time, because of our haste, we were carrying a $60,000 professional camera, so people knew we meant business. This magnet also attracts the people who throw their eight year olds at you and tell you to make them a star. Everyone also believes their baby is the cutest on the planet; several people are very wrong in their perception. Figure 24-1 shows a happy couple who were guests and just happened to act in our travel video.

Don't get me wrong; I'm not telling you to buy cards, shirts, jackets, and an expensive camera to show off to people and hang out at the mall to prove that you are a professional. These are just tools that may make this role a little less difficult or awkward for some people to swallow, allowing you to get people to appear on camera. I dislike people who pretend they are something they're not (I don't mean actors). The only thing you should have when you approach strangers is a business card and an actual need for them in a video. It should list your phone number, name, and other information so they can contact you. Don't hand-write your name on a piece of paper if you truly believe you are actually going to get someone.

One last thing about this type of actor pickup: After the project is finished, make sure you pay them a few dollars for their time or ask the client if you can give them a sample product. This little gesture goes a long way; it makes you, your company, and the client look good. I have friends that made a quick $50 for just spending an hour with us. Most budgets are large enough to accommodate that. But if you really have no money, at least offer them a copy of the finished tape.

Figure 24-1

Nonprofessional actors working for next to nothing

#201

We had a gentleman drive his classic 1957 Thunderbird past the camera, covering three seconds of screen time in a corporate video. The client wouldn't give him a cent, but I gave him a copy of the 20-minute video and he was speechless. Not only could I rely on him again for a another video, but he told his classic car friends that he was now "famous" and showed the tape to them. Consequently, I now have access to his entire club of 40 classic cars for use in my productions. A cheap $1.50 dub netted me 40 new contacts. Figure 24-2 shows the classic T-Bird on location.

One last thing about the mall or any public place: What happens if you approach someone, they respond favorably, and you find out once they are in front of the camera that they cannot act, take screen direction, or basically just don't work out? They may be trying their best, but some people just don't have what it takes. The best theory in my opinion is to thank them, give them something for their time (besides a pat on the head), and send them away. The first thing they will ask is, "When will I be on TV?" I should tell them "Never!" but instead I explain that not everything you shoot makes the final program. Everyone has heard of the cutting room floor, although it's all electronic now. Tell them it's not your decision (it really isn't—it's the client's), but you'll do all you can. This is not a lie; you will do all you can . . . to keep them from appearing in the video.

Figure 24-2

A '57 T-Bird in its natural environment

#202 How to Use Multiple Crews and Get the Same Results

Sometimes you're handed a project that seems almost impossible from the start. The client wants it yesterday; you have miles of footage; and he, she, or it wants total approval of each process. But the hard part is the client wants you in 12 different places at once.

I once had to do 12 single-person interviews, shot in 12 different cities, in the course of 1 week. I had to hire crews in each of the locations to perform the task of shooting. Of my newly assembled dirty dozen, I wasn't able to get every crew to shoot in the same format. I had crews shooting in Betacam SP, Betacam SX, DVCAM, and DVC-PRO. The format the crews shot in wasn't important; consistency would arrive later. Our main objective was to have all the footage look somewhat similar so it could be easily edited together.

A few steps can be taken to keep things looking the same. Create a detailed list of what each crew should do so in preparation and shooting, so the project looks like the same crew traveled the globe. Also devise a "form letter" that details what you expect of every crew. No detail should be left to their imagination if you want all the footage to look the same. Our crew in China matched that of England, South America, and the States. (Of course, the footage from overseas was shot in PAL, so the standard had to be converted to NTSC.)

Since we were receiving four different tape formats and two different standards (PAL and NTSC), we transferred everything to the format of choice. This was where consistency was the most important. If you try to digitize a DVC-PRO tape directly into your editing system, it won't match the characteristics (good or bad) if you chose Betacam SP as your standard. Transfer the master tape to Betacam first, and then digitize it. As long as the field tapes are all transferred to the same format and standard when editing begins, the differences will be slight. If all your crews follow your roadmaps, the footage will be the same.

#203 How to Find Narrators and Voiceover Talent

Unfortunately, there really aren't any bars or restaurants where men and women with great vocal cords hang out; you have to find them elsewhere. Two types of vocal professionals can be used: union and nonunion.

#204 Using Union Talent

Union talent, as the name implies, is someone who makes his or her living speaking. Their fees may be higher than nonunion, but that doesn't necessarily make them any better than narrators who don't belong to their organization. More is involved when working with union talent and some of the headaches you may encounter with nonunion people are eliminated.

You will have access to a nationwide chain of men and women whose voices may perfectly lend themselves to your video production. Contact both the Screen Actors Guild (SAG) at www.sag.org and the American Federation of Television and Radio Artists (AFTRA) at www.aftra.org to get information on obtaining narrators.

#205 Using Nonunion Talent

These folks, also extremely talented, can be found locally for your video productions. I've used local radio and television personalities in my area to supply my vocal recording needs. These people all pass the test of being able to read copy and speak clearly into a microphone.

Check local theater groups and acting workshops, as well as college students majoring in communications, to get a greater range of male and female voices. However, remember that the more professional choices are easier to work with because they may need less coaching and training from you.

Lastly, don't just stop someone on the street because he or she has a great voice. It does take skill and training to master the art of narration. I can read copy better than most because I have been in this field so long, but my voice doesn't have the quality needed for narration.

Solutions in this Chapter:

Chapter 25

Special Effects

Stage Blood, Squibs, and Makeup

#206 How to Create Special Effects

Cheap special effects are one way to lower the production costs of your digital video. Your job is to spend as little as possible to make your effects believable but also not call attention to that fact.

In this chapter, we will talk about creating stage blood that looks real and tastes good, exploding squibs that propel blood and tissue to simulate a gunshot wound, after-the-fact wounds that look painful, dead bodies that pass the test, realistic ghosts that don't look like white sheets, and making your actors look younger or older than they really are.

#207 How to Make Cheap Blood

If you ever need an exploding blood squib and don't have the money, expertise, or contacts in the business (not the Mafia), look no further than right here. Although you are not using gunpowder in these squibs, the mixture is flammable, so keep your little brother out of the room when you make it.

Stage Blood Formula

4 parts Karo® clear corn syrup or equivalent

2 parts chocolate syrup, Hershey's® or equivalent

1 part red food coloring, Durkee® or equivalent

1 part water

1 drop blue food coloring per 59 cc (1/4 cup)

Mix up the blood in a condom – yes, that's what I said. Put your exploding mixture into the condom and tie it up. You now have an exploding blood squib. Attach a 25-foot length of lamp cord to the leads sticking out of the squib and attach the other end to a car battery. Twelve volts is needed to safely rupture this sticky, staining mess. My advice is to test at least five of these before they are attached to humans.

#208 How to Make Exploding Blood Squibs

You've asked for it and you've seen it in the movies. Now it's time to add these effects to your digital video.

Recipe for Exploding Blood Squibs

Two five-inch pieces of .51mm insulated copper telephone wire (24 AWG)

One single-strand copper wire 1/2" in length (18 AWG)

One-pound jar of Pyrodex®

One disposable surgical glove or condom

One 25-foot length of lamp cord (18 AWG)

One four-inch-square piece of 18-gauge steel

One car battery

#208

The igniter is the element that makes the blood-filled bag blow up. Cut two five-inch pieces of insulated copper telephone wire .51mm in diameter (24 AWG). This will be called the lead wire. Strip one inch of insulation from each end of both wires. Cross the wires at 90 degrees about a quarter of an inch from the ends. Pinch the long ends of the wires in one hand. With your other hand, twist and pull the short ends of the wires until you have about a half-inch of wire twisted together in a tight, rope-like spiral with a small fork at the end of the twisted section. Make several of these units.

Find someone with extremely good eyesight for the next step. Stretch one strand of copper wire from the left fork prong to the right, wrapping the wire around this prong five times. The fork's cap should be only 1/16 of an inch. This single strand of wire can be from an 18 AWG lamp cord.

Now that you have the igniters, you need to ground the powder. Purchase a one-pound jar of Pyrodex from Wal-Mart or any sporting goods store. Pyrodex is a muzzle-loading propellant that is smokeless and fireless if used correctly. Because the powder is granulated, you must crush it into a fine, soft powder. If you don't, the powder will ignite and smoke.

Figure 25-1

An exploded squib on a body

Pour a half inch (no more) of Pyrodex into the cutoff finger of a disposable surgical glove and stick your igniter into the powder so the single-strand lead is buried. I know the finger isn't filled, but you need air space for the powder to ignite and burn effectively. Tie off the finger securely so it is blood- and airtight. Place the finger inside the blood mixture filled condom.

Once we were sure of ourselves, we attached our squib to a 4-inch-square piece of 18-gauge steel. This steel was what protected our victim, I mean talent, from the charge. Use gaffer tape to wrap the steel before you tape the squib to the plate. Leave a large section of the condom exposed (stop laughing) in the front so the blood can rupture. Have your actor wear a T-shirt and tape this steel-plated squib to his chest.

The charge isn't strong enough to rip the fabric of the shirt, so cut a small X in the shirt directly over the squib. Leave one strand of fabric still attached so the shirt looks intact; the charge will break though the condom and the weakened fabric. Figure 25-1 shows how realistic this effect can be.

#209 How to Make Realistic-Looking Wounds

Because video is such an up-close and intimate medium, the quality of your special effects must look good or the viewer will see right through them. Getting wounds to look real takes a little skill and the right kind of makeup.

It mainly depends on the type of wounds you're looking for: puncture wounds, bleeding wounds, or missing appendages. Obviously, the level of difficulty increases with the complexity of the wound. Let's talk about the easiest first.

#210 Puncture Wounds

These are the wounds left when someone is stabbed or shot; something penetrates the body and leaves its nasty mark. Blood will flow from these wounds soon after they are inflicted, but if you desire to show a "healing" wound, look at images of the wound in books, TV, or on the Internet.

The key thing in most of these wounds is discoloration. Rarely is skin all clumped up on one side or the other, but we will discuss that effect shortly. Determine the hue of the wound; a dark, bluish-black color means blood has collected under the skin, while a yellow color is the same wound in a nearly healed stage. Use different colored makeup to get the correct color combination for the "sore" part of the injury. Rub the area with your finger to smooth and blend the colors.

To create a ridge of dead skin or a scab, take an adhesive (*not* superglue) and put it on the skin. Hold the skin together for a few moments so it dries in a bunched-up position. The large scab can be made with flesh-colored putty and attached to the skin the same way or with spirit gum.

#211 Bleeding Wounds

To make a wound more "active," blood should be oozing out. To show a close-up of a stab wound, dress a sack in the actor's clothing, simulating the correct part of the body. Put a baggie of your stage blood under the stab area. Shoot a close-up of the knife going into the clothing (on the sack and blood bag). Blood will now flow out as if it were a real wound.

Flowing blood can be supplied via a tube and a pump as long as they are on the opposite side of the limb and cannot be seen by the camera. By soaking tissues or putty in your stage blood solution, that too may be applied to the wound.

Add a glossy finish (sounds like a furniture commercial) to the scab to make it appear more fresh and recent. Glue that dries shiny also will do the trick when painted on the sore. Any little bits of debris and tissue fragments should be added to intensify the wounded area.

#212 Missing Appendages

This is the most severe of special wound effects and is best not to dwell on this in a close-up. Removing a limb and leaving a bloody or spurting stump can be done in two ways. The first is to do it the Hollywood way (electronically) and the other is the old-fashioned way.

By draping the appendage in Chroma Key fabric, you can electronically remove it in postproduction. Various methods are available for doing this and, depending on your software program and skill, this may or may not be difficult.

The other way is to cover a "to-be-missing" hand with a stocking and more stage blood. Little things like getting a longer shirt sleeve, cutting it, and hiding the hand in the bloody stocking (sounds British) will give the appearance that one hand is intact and the other is missing.

In a recent PSA for safety, I had an actor reach into a snowblower to remove a clog. Since the machine was running (we made it appear if it were), his hand was mangled by the auger. This was a very effective shot because of the shock value. Use this to your best advantage by having the talent shriek in pain. This sound catches the viewer off guard for a moment and it causes them to blink or lose focus for a millisecond.

Immediately cut to a close-up of the talent's face. The viewer will not want to see this, but it intensifies the action. Then cut to the mangled hand. This was very easy to show because an auger does not cut cleanly. Instead, it twists and mutilates.

I had the actor hold a piece of flesh-colored cloth with blood dripping from it to serve as the damaged hand. The cloth was actually a stocking again with pieces of putty (chopped fingers) and blood-soaked tissues inside. Your audience should not be able to dwell on this for a long time. If this were to really happen, the actor would not be able to hold it or look at it for long before he or she passed out. Only give the audience a split second to see this and keep the hand moving. Therefore, they have no time to focus on what they are really seeing and the effect works.

In cases like this, the less you show, the better and stronger the effect. Unless you are doing a schlocky, cheap horror video and want to call attention to the fact, show them very little and leave them wanting more.

#213 How to Create Dead Bodies

A time comes in everyone's life when they must try to videotape someone who is supposed to be dead. You've all tried to get the talent to hold their breath so the viewer can't see their diaphragms moving up and down. This works in a pinch, but any subtle movement, twitch, or flinch destroys the impression.

Another approach is to use a dummy or mannequin as the dead body. Mannequins, although somewhat poseable, never really look like a human with their unrealistic sheen. Depending on your budget and the time you have, preparing a dummy is very popular if you have the artistic ability to sculpt latex rubber and paint the flesh the correct hue. Once again, you don't want to dwell on this shot because if the viewer sees it for too long, the effect is lost.

My favorite approach is to show the dead body from a distance – briefly. Any exposed flesh on your "corpse" should have makeup applied to simulate a deceased appearance: pale, clammy, and slightly blue. Prepare a mixture of white pancake makeup and a slight amount of blue. When mixed together, you want the color to resemble gray. When a human dies, the blood leaves and all that remains (a dead joke) is pale, lifeless skin.

Apply this makeup lightly to the face, hands, and any other exposed areas (if the body's feet are bare, apply makeup there also). Spread the makeup thin enough that it doesn't look caked, and a slight amount of black should be under the eye sockets (a slight amount). Apply wax (mortician's wax is best) to your victim's lips to take the shine and lifelike appearance from them.

Pose the body with his arms down at his side if doing an autopsy scene or bend them if the victim has been in an accident. If bones are supposed to be broken, the body should look that way. This is the best time to use dummy limbs because they will twist and contort easier. If the body is on the ground and a leg is broken, dig a hole for the real leg and have your talent stick his leg through a hole in the bottom of his pants into the ground. Then stuff the remainder of the pant leg with rags and stick on a shoe (with a stick inside to hold it in place). Smooth out any lumps and twist this into the desired position.

If doing close-ups on the dead body, try to avoid showing the face because this is the most difficult to make look dead. The eyes will be white and lifeless (white contact lenses), the lips cracked and broken (pieces of wax applied to the lips), and the wounds bloodless, as described in the last section. Instead, just add makeup to the arm or leg of the dead being and go from there.

Always try the simplest solution first before digging yourself into a hole (unless that's where the broken appendage is being placed).

#214 How to Create Realistic Ghosts

We've come a long way from people wearing white sheets pretending to be ghosts. Instead, most insert these spectres digitally via a software program. This is the best solution, but also the most expensive. A cheap route is the old-fashioned beam splitter.

A beam splitter is a partially silvered mirror that reflects up to 80 percent of the light while allowing 20 percent to pass through. With a beam splitter in front of your camera, a transparent ghost can be seen because some of the room light will pass through it. The effect was devised in the 1930s and it still works today.

These semi-mirrors can be purchased from catalogues or the Internet at www.berezin.com/3d/beamsplitterinstr.htm. A full mirror won't work because all the light hitting its surface is reflected back. With a partially silvered or reflective mirror, just the right amount for a ghost is reflected.

Have your ghost stand in an area to the left or right, depending where you want them to appear in the frame. The ghost's background should be black because you don't want any other reflections to be transparent (other than the ghost). While shooting through this splitter, light the ghost and look through the viewfinder until your spirit is illuminated or ghost-like enough. The more light that is applied, the brighter and less see-through the ghost is; less light means more transparent.

Light the rest of the scene as you normally would, leaving a space in the frame where you want the ghost to appear. Shoot through the beam splitter and the scene in front appears normal. When you apply lights to your ghost, it will materialize in the frame because of the silvered mirror. Fading the ghost's light in and out makes the ghost do the same onscreen. Practice your timing and position of the ghost and you now have a realistic effect.

Since most ghosts wear white, this will be the first thing visible in the frame (the white outfit, then skin, and so on). Using a fan, eerie makeup, pieces of tattered cloth, and bluish lighting, your ghost will look just like the real thing (I see them all the time and that's how they look).

#215 How to Make Your Actors Look Older

It's always best to get an actor who is the same age as the role, but this isn't always possible. Times will occur when a younger person must look older or vice versa, yet this can be accomplished through makeup.

Let's begin with hair. Hair can be aged gracefully in two ways: One is with gray hairspray and the other is with washable shoe polish. Baby powder or flour works, but any movement will send puffs of white into the air. Unless your senior citizen is dissolving, don't use powders. Hairspray is basically paint that tints the hair gray, and the more you use, the grayer the hair. Liquid shoe polish is messier but tends to stay where you put it without hardening the hair like spray does. Once the hair is the correct shade of "older," focus on the face.

Makeup is the best way of aging, with latex wrinkles being added for more advanced years. Like creating a wound, use a nonpermanent adhesive and pinch the skin together to create crow's feet around the eyes. Never put anything too close to the eye! Bags around the eye sockets, dull, pale skin pallor, and wrinkles are what creates age.

The talent should move and speak slower, and their clothing should reflect their age. A young person playing someone very old should wear baggy clothing, simulating the shrinking process associated with aging. The pants are worn higher on a male, and stockings can be bunchy on a woman.

The amount of care taken in the makeup is the key to believability. Look at close-up shots of older people and try to duplicate that on your talent with makeup, clothing, and accessories. The last step is rehearsal. If the actor doesn't move and sound convincing, the best makeup in the world won't help.

#216 How to Make Your Actors Look Younger

If you want to show an older person in their younger years, you have to perform the fountain of youth ritual on them. This involves deception and isn't too difficult to pull off.

Begin with the hair again and let the actor comb it down or basically make it look longer. Younger people are into bangs, and older people wear their hair up or pulled back off the face. Use a wig on balding people and hair coloring to give it a more youthful appearance. Blonde hair also signifies youth and works on both sexes. Removing facial hair on older males can help remove years.

The face is slightly more difficult and you can use the same process of make-up for aging. Instead of bunching the skin, pull it tighter toward the ears and use spirit gum to soften and smooth out the wrinkles, much like what happens in a facelift. Makeup should be applied, with brighter flesh tones being the color choice. Women should also use brighter colors in lipstick and mascara.

Clothing should be more form fitting to accentuate the shape of the body. Movement and speech should be slightly faster and more deliberate, implying youth. Lighting should be softer with more diffusion and filtration. If your camera has a detail circuit for flesh tones, choose a soft setting. At least now they will appear younger, if only on camera.

Chapter 26

Action Videos

#217 How to Create and Record Explosions

As a kid, you would always get into trouble if you tried to do something like this, but as an adult and in the course of making a video, it's okay.

Two types of explosions exist: the dangerous ones and the not-so-dangerous ones. The first grouping involves volatile explosives like dynamite and gasoline. These pyrotechnics should be left to the experts and will not be covered in this book. The less obnoxious explosions still look great but are slightly safer. Anytime you are working with an explosive, use extreme caution and keep others who aren't involved away.

If you desire an exploding fireball coming up from the ground like a bomb on impact, you need a small charge of Pyrodex (mentioned in Solution #208) and some white flour. Dig a hole about one foot deep and fill a third of it with flour. Fill the rest of the void with shredded pieces of debris like leaves, paper, and anything else you want to blow out with the blast.

Insert an exploding squib (review Solution #208) into your hole (on the bottom) and cover the wires. Use a slight bit more Pyrodex for this exploding squib and grind it less if you want smoke also. The pros use this same technique but with a stronger charge (this charge is powerful enough for our purposes). On cue, touch the positive wire to your battery and the blast will occur, igniting the fine flour. As the tiny mushroom cloud vanishes, the debris will filter back to earth. Always use caution when using Pyrodex and make sure your talent isn't close to the explosion when it goes off. Firecrackers will be more impressive with the explosion, but I do not recommend them because you cannot time the blast when lighting a fuse.

The sound this blast makes is pitiful so it must be enhanced in editing. Sound effect libraries have great explosions and theirs will be far greater than anything you could record on your own.

If you must record an actual explosion for a video, do not use a microphone that may be damaged by the shock wave and blast. If you are far enough away from the blast, a shotgun is your best choice. Just remember the volume of the noise will be high, so plan accordingly.

I had to record an explosion for a training video and the microphone had to be close to the source. The explosion would destroy a quality microphone, so I used a cheap $5 model. The client insisted that I capture the actual noise this blast made and he paid for the cheap microphone. I got the sound right before the mike was caked in soot. Cheap microphones are tougher to destroy and can handle explosive sounds better than ribbon or condenser mikes.

Once again, lots of caution is the most important thing to remember. If you aren't comfortable with doing explosions, my best advice is, don't do it!

#218 How to Choreograph and Shoot Fight Scenes

All fights scenes must be planned out in every detail so no one gets hurt in the process. Besides, staged fights (like those in wrestling) always look better.

Every movement, swinging of a fist, slashing of a sword, or throwing of a bottle, should be rehearsed so the actors and camera operator know where to be at a given time. This does take time and is worth the effort. No ad-libbing should be allowed in this process because it will only lead to disaster.

At all costs, fight sequences should be videotaped relatively close. No one wants to see people fighting from a distance. If the viewer is up close and personal with the fighters, they become more involved with the characters. We need to see people's faces so we can react to their pain. Subjective shots are a great way for the viewer to identify with someone in the fight. Don't take it as far as the camera getting punched, but a swing toward the camera and an immediate cut to the actor reeling from the impact is quite effective.

The key is for actors to get as close to hitting their opponent as possible without actually making contact. If character A throws a punch at character B, B needs to fling his head back as the fist comes in. This is where timing and the camera angle are crucial. If the camera can see a large gap between the fist and the face, move closer to the action to close up the gap.

A low-angle camera should be used to make the fighter appear more dominant in the frame, and a high-angle shot to show the opposite. The hero is shot from below and the antagonist from above. Every time you see Steven Seagal, Jean-Claude, or Arnold in a fight, they always look more imposing because they are filmed from below.

Sound effects are what make fight scenes more "video realistic." In real life, punches don't sound like they do in movies (neither are gunshots for that matter). Since we are attuned to hearing them in this genre, if they are absent we think the fight is phony.

#219 How to Shoot Car Chases

One of the most exciting aspects of an action video is the car chase. Made famous with Steve McQueen's *Bullitt* (although car chases have been around since the early 1900s) and taken to the extreme in films like *The Fast and Furious*, the chase gets everyone's adrenalin pumping.

Like a fight scene, car chases must also be choreographed in great detail, knowing where each car will be at a given time and where the camera will be to capture them. You will be manipulating expensive, heavy machines that can injure as well as kill, so precautions need to be taken. Here is where fast cuts and subjective camera shots work extremely well.

Map out the area where the chase is to happen. The roads should be cleared of any vehicles or people that should not be there. The term liability should come into play because you are dealing with property and cars.

First start with an establishing shot showing which cars are involved. The audience needs to know who is driving what kind of car. Throughout the chase sequence, cut (not dissolve) between interiors of the drivers and exteriors of the cars. In interior shots, showing shifting, steering, braking, and so on allows the viewer to identify with the driver. Immediately cutting to a tire screeching around a turn after an interior close-up of the hero turning the steering wheel suggests to the audience a completed action. They will put two and two together even if you shot out of sequence.

This is another reason why storyboards (choreographing) are important because all the exteriors, interiors, and subjective shots are shot together. You don't want to be moving the camera 300 times; instead shoot all the similar scenes together and assemble them in the proper order in post (following your storyboard).

Subjective shots add to the excitement by putting the viewer in the driver's seat. Everyone has wanted to be in a car chase but didn't want to wreck their car or incur a thousand tickets. By attaching the camera to the front fender and getting the tire in the shot, this adds movement to the action. A subjective camera shot from the hood of the car, the roof, and even the trunk gives a point of view that allows the editor more choices in the cutting process. Also, get a point of view shot from over the driver's shoulder so we can see his hands on the wheel and react to what he's seeing out of the windshield.

Once the camera is mounted on the car, you don't want to move it after every shot. This is why shooting out of sequence is so important and time saving. The sound effects (always amplified) and fast cutting quicken the pace of this visual feast. Don't dwell on any shot, no matter how exciting, for too long. This bores the viewer and actually slows the pacing of the chase.

To add a slightly faster pace to the scene, I've had drivers keep their speed down and played the video back at an accelerated pace (115 percent). Don't speed the action up too much or it looks silly. Once again, make sure you have clearance from the proper authorities before you just go out and shoot it.

The object is for the cars to get as close to each other (and their surroundings) without actually making contact (just like in a fight sequence). These close calls are even more thrilling than the collisions to some because it means the drivers have greater skill in handling their vehicles.

If you have the budget and the okay from the police, actual slight crashes offer even more realism. This rarely happens without larger budgets and duplicate cars. Every feature film with a car chase has several (sometimes three or more) identical stunt vehicles to be used for the chase sequence. When car 1 gets hit on the left side, car 2 (used for the close-ups) has to have that same dent on the same side for the audience to buy it. If not, they know that the cars have been switched.

In the *Smokey and the Bandit* series, the Trans Am would leap off a bridge, crash nose first into the road, and then drive away after the camera cut. Obviously, the car was totaled in the crash because we saw the front end crumple. When the camera cuts to another version of the same vehicle, we usually buy it because we want to see the Bandit get away with it. We saw the car destroyed, but the car can do almost anything in a movie.

We've also seen hub caps fly off as the car careens around a bend. The next cut should have that hub cap missing from the same wheel or the viewer will notice a continuity problem.

Take your time planning the sequence and decide on paper what looks best and when from each angle. The finished video can be cut in a multitude of ways, but you need to know beforehand where each vehicle is going to be. Watch chase sequences and see how the shots and pacing work for them. Just make sure you have permission, the skills necessary, and the insurance before you grab a car and begin the chase.

#220 How to Shoot Action Sequences

Action sequences involve close calls, fast cutting, and a slight bit of believability to bring the nearly impossible, staged action to life.

Cutting is the key to any action video and a good editor needs lots of camera angles from which to choose. If the camera is stationary and enables the action to occur without close-ups, reactions, or subjective shots, you have a no-action video.

For each action to happen, you must first establish what the situation is for the viewer: The hero is in the wrong place at the wrong time, the female lead must break away from her captors, or seconds remain before the bomb explodes. With the first, long, establishing shot, the viewer needs to know what the situation is and where the trouble might be. This gets them thinking of a solution, which is important.

As the story progresses, the shots are cut shorter and shorter to quicken the pace of the video. You should cut from long shot to close-up for character identification (sweat on the brow, quivering lips, tears in the eyes). Cut again to establish the problem ("I'm shot and blood is flowing," "My leg got blown off," or the train is about to run over the heroine) to a reaction shot of the hero ("How am I going to get out of this?"). Then cut to a subjective shot that shows the possible solution to the problem ("there's a window—if I could just get across that field of glass and alligators") to finally an attempt to solve the problem.

If the first attempt fails, which it should, that will build the suspense and get the viewer to think (subconsciously) of other ways out. This process may be repeated over and over until the situation has been solved. It's not so much the process; it's how the hero gets there that makes an action video work.

#221 How to Shoot Martial Arts

Like a fight sequence, a martial arts video involves the same choreography only with legs, feet, exotic weapons, acrobatics, and incredible odds (20 people fighting 2).

Every single move must be planned because an amateur could really get hurt. Always use someone who has training to choreograph this sequence, especially when using weapons. Here again, the sound effects are very important. When someone gets hit over the back of the head with a chair, the viewer wants to hear the wood as well as bones breaking.

Even though bare feet are often used in martial arts, the audience doesn't want to hear dull thuds. The sound must be enhanced. I've recorded celery sticks being snapped to resemble bones breaking, feet kicking burlap sacks filled with peanuts for body contact, and slapping a belt against a table for fists against flesh.

Two other things are important in choreographing martial arts videos: slow motion and fluids escaping. Slow motion shows the viewer more clearly what is happening in the fight and how acrobatic the fighter must be. Back flips over people's heads, walking up walls, and other unnatural moves necessitate wires on the actors and people off camera to help them into the air. This is complicated and involves wire removal in post. Instead, use punches and kicks in slow motion.

Have the actor doing the punch or kick perform the motion at normal speed with the other party not in the shot. Play this shot back in slow motion in editing. Cut immediately to a close-up of the other person receiving the blow. Since you are shooting this in close-up, just have the fist or foot move three inches away from making contact. This will not hurt your actor because the distance and velocity of the blow are minimal. This close-up is where the viewer needs to see actual contact. The recipient should then react to the impact. This should also be in slow motion if the punch or kick is in slow motion. This is one of the few times not to utilize a moving camera. If the camera is shaking too much, much of the action may be lost.

A handheld camera is fine for a fight like this because it is more fluid and the slight wobble looks realistic. If the camera is locked down on a tripod, the actors must perform within a given space and that may tax their abilities. Dolly moves or Steadicam shots can arc around the fighters and give different perspectives of the fight.

Having saliva fly out of the mouth or sweat from the body escape on contact with the blow makes any fight seem more vivid. Having the actor keep a mouth full of water (it looks like saliva) or spraying an actor down with water will simulate the effect nicely.

Chapter 27

Horror Videos

#222 How to Shoot Horror

Horror has been given a bad rap lately because it seems most people want to shoot horror because the budget is low and they feel "gore" is cheap. A lot of great horror films are scary because of the story, however, not because of the special effects.

Creating horror in a movie is achieved by making the viewer feel what the director wants them to feel. The audience needs to ask, "What's behind that door?" "I wouldn't go in there if I were you!" or "What does it look like?"

As in any genre, anything looks bigger and scarier if shot from below. Little monsters aren't as frightening, so why not make them look bigger? The same applies with quantity. If 1 dinosaur scares you, 63 will make you jump out of your skin. With the camera, you control what you want the viewer to see. Sometimes even the color can create an impression; black and white or muted colors are scarier than vibrant, happy colors.

In my horror film, *Nobody*, the heroine is trying to run away from the apparition chasing her. When she wanders into camera frame, she thinks she has eluded her pursuer. As she cautiously walks forward while looking back, I panned the camera with her, left to right, letting the frame lead her slightly. In other words, instead of her being centered in the frame, the camera pans ahead of her, leaving her toward the rear of the frame. The creature is standing there as the camera pans with her. Still looking back, she bumps into him. The camera showed him a split second before she saw him, so the viewer screams when she does because both notice him at the same time. Figure 27-1 illustrates the framing of a sequence like this.

Remember, the imagination can come up with something far scarier than the camera may reveal. That's why reading a horror novel is always more frightening that seeing the movie version. You were terrified when you read the book, yet the monster in the

Figure 27-1

Watch where you're going.

film is not quite as scary as you imagined. As soon as the camera reveals what's scaring us, we think, "That's not so bad!"

As a digital filmmaker, you must scare the viewer with what you show or, better yet, with what you don't show (we'll discuss this more in detail in Solution #226). When showing the audience something, it must elicit a response. The shorter something is shown onscreen, the more frightening it becomes. A brief glimpse is better than a lingering view.

A handheld camera is shakier and scarier than a tripod-mounted shot because it makes the viewer uneasy. The same is true with a Dutch tilt (when the camera is skewed to the left or right and not straight on). Anything that is away from the norm or unconventional is good. The object is to throw the viewer off guard; when they least expect something, that's when you hit them between the eyes.

#223 How to Light for Horror

Many people think that you don't light for horror; the opposite is true. Lighting for horror videos is extremely precise because you must know where every shadow is going to fall to give the impression that something evil is lurking in those shadows.

Low key lighting is what makes horror stand alone. The shadows are deep, dark, and foreboding because the lighting instruments are placed low and shoot up, distorting anything in their path. Instead of high key, bright, modeled three-point lighting, often just one light, a key light, is used without the need for a fill light. That means one side of the face or person is lit and the other side, without an additional light, is bathed in deep shadow.

The lighting is rarely soft and diffused (unless used to throw the viewer off); instead, it's hard and directional. Sometimes a practical on the set is the only light. Horror needs shadows and the starkness and an open bulb does the trick. Look at the fabulous lighting in Alfred Hitchcock's *Psycho*. When Vera Miles discovers Mrs. Bates for the first time, the only illumination in the room is from an overhead bulb on a frayed cord. The bare bulb casts long shadows from above, making eye sockets look deeper and longer.

When Vera grabs Mrs. Bates's shoulder and swings her toward the camera, discovering she is a well-preserved corpse, she reels back in terror and hits the overhead lamp with her hand, sending it swinging back and forth. With the addition of screeching violin strings and Vera's screams, the swaying light plays tricks with the shadows as they switch between illuminating and enveloping the corpse in darkness.

As a child, you held a flashlight under your chin in the dark to make your face look ghastly. That is the same principle you use in horror; you do not want flattering lighting (which lighting from below never is).

In my horror film *Nobody*, I rarely used more than one light in any scene. As I mentioned earlier, the deep shadows were always hiding something, whether intended or not. Figure 27-2 illustrates a low key lighting approach.

In no other genre of production can you have as much fun with lighting as horror. Fans may be used to create mysterious wind, blue lighting is accepted as natural (or use daylight balance with tungsten lights for a colder appearance), and darker exposures (underexposing) are best.

Try shooting your horror video in black and white. The chroma gain on your video image can be lowered in editing, and shooting through a red filter will deepen all blacks and make flesh tones very white and pasty. One thing to avoid in horror videos is video noise (grain). The lower the light level, the grainier the image. *The Blair Witch Project* used low light and grain beyond acceptability and made the viewer uncomfortable.

Instead, light your set for an average exposure of F5.6 or F8, but underexpose when you shoot (set the camera at F11 or F16). This way you get underexposure without the grain in the blacks. If your set is lit for F1.2 and your camera only opens to F2, that's where you get grain. Since you are in control, light slightly brighter, but stop down for underexposure. This will help scare the viewer even more.

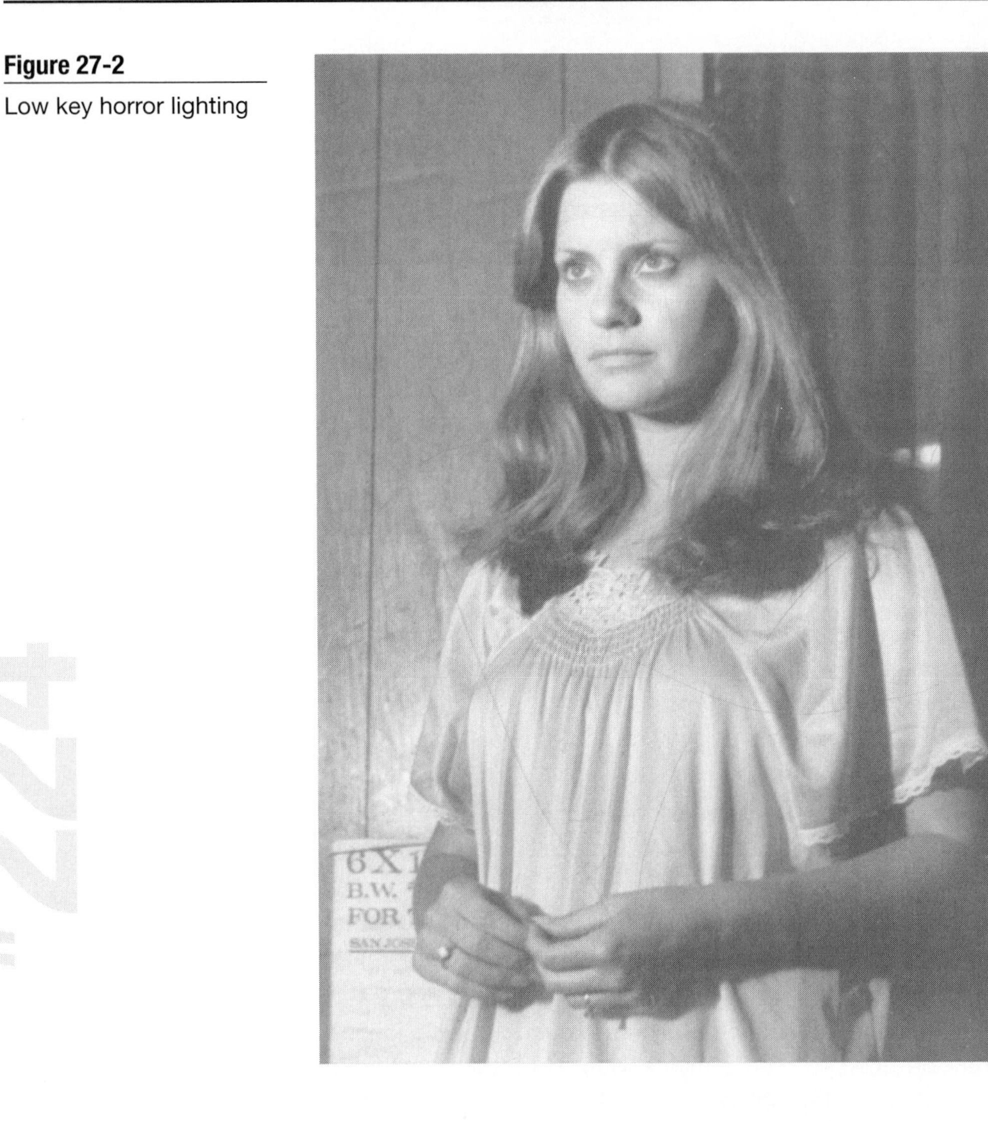

Figure 27-2

Low key horror lighting

#224 How to Edit for Horror

Editing establishes the pace of any video, and horror is no different. You can do some tricks in horror, however, to make the viewer scream.

Begin with an average pace in editing, cutting where you normally would in any video. When you want the tension to mount, increase the frequency of cuts (not to the pace of a montage). With faster pacing, the viewers are getting as scared as the talent. They have no idea what to expect when and if something is going to jump out at them.

A great trick used in the original version of *The Thing* (1951) was when the group or stranded scientists thought they had the Thing cornered.

We've all seen this before, such as when someone reaches for a door and throws it open. We always expect something to be in there because the music is building and it crescendos when the door is open, revealing nothing inside. This device has been done for years and it still works. In *The Thing*, they open the door expecting to see something, and *it is* occupied by James Arness (the Thing). Since the viewer didn't expect to see anything in the closet (because they never do), the editor fooled them by putting something *in* the closet.

A fast cut to the monster as it snarls works best in a two-second shot. It's never long enough to see detail, but it establishes the creature and makes us uncomfortable. Any image onscreen for more than 11 frames is registered as being "seen" consciously. Subliminal advertising used to work because words or phrases were shown to the viewer for only one frame. The eye notices a little blip on the screen, but the brain registers and reads the image ("remember, buy popcorn!").

Use this to your advantage and only put something onscreen long enough for it to be registered in the viewer's mind without lingering too long. Use lots of point-of-view shots, speed the pace of editing when you want the audience's pulse to beat faster, and slow it down when you want them to relax.

You want to scare the viewer often with your shots and editing, but they also need time to relax after a series of "scares." If a comedy never lets up, the viewer is exhausted when leaving the theatre. After you've scared the pants off someone, give them a calm, slowly paced scene to let them catch their breath (then hit them again when they don't expect it). Remember the last scene in *Carrie* when the camera slowly dollies through the cemetery and we watch Nancy Allen put flowers on Carrie's grave. After all the carnage that happened at the prom, the audience is still wound up but slowly calming down. Suddenly, Carrie's bloody hand reaches up through the grave dirt and grabs Nancy's hand. The audience never expects this because everything was over, and the director got in another unexpected scare—through editing. After the scare occurs, viewers like to turn to each other to see their reaction or discuss what just happened. If you do this nonstop up until the climax, scaring the viewer will get no relief.

When I saw John Carpenter's *Halloween* in the theatres in 1978, after a series of scares had occurred, leaving the viewers with a bad case of jangled nerves, some joker in the audience decided to slam the exit door after they returned from the bathroom. The crash of the door made everyone in the theatre scream (even me) at the sudden noise and then laugh afterwards when they discovered the source of the sound.

In the movie *Stir of Echoes*, Kevin Bacon has been seeing and hearing things. In one terrifying shot, he looks at the sofa and sees a young girl with a white complexion, black eyes, blue lips, and a plastic bag tied around her head. This chilling apparition is only onscreen for one second, but the suddenness, the repulsion of what has happened to this young girl, and the surge of music terrify us.

In the same movie, Kevin is standing in the bathroom looking in the medicine cabinet for an aspirin. When he closes the mirror, we see the same pale spirit staring back at him. It's a brief shot, we don't expect it (neither does he), the music blares, and when he reacts and turns around, she's not there.

With editing, you need to get your viewers agitated the same way, so any sudden noise or bump in the night will make them scream out. You've built them up for something; now use it to your best advantage.

#225 How to Use Sound Effects to Scare

Sound is often used as an aid to enhance the visual onscreen. What the viewer has just seen is scary enough, but adding a "stinger" sound effect (a sudden burst of sound or music) intensifies the visual. With all the senses heightened when watching a horror video (the visual is well taken care of), use the aural function to add to the impact. Eerie music puts us on edge, but a sudden sound or noise gives an additional boost.

I made the mistake when watching a horror movie of trying to get my date to hand me the popcorn. Instead of asking her for it, I touched her hand. Since her sense of touch was heightened, I was showered with popcorn (as were several others) when she shrieked at my touch.

Sound effects can also spark reactions in the audience. If one person yells out when she is scared, that yelling "sound effect" helps your cause. Any sudden burst of sound scares people. The key is to time this outburst.

In editing, a sound effect should be placed a few frames before the scare begins. Immediately upon hearing the sound effect, the viewer will react (and then see the scare onscreen). A sound effect can be any noise; it's just the suddenness and intensity (make it loud enough) that frightens. Sound effects can be intensified to create uneasiness, such as adding a heartbeat to the soundtrack. When you speed this beating to a faster pace, the viewer's pulse also quickens. Discordant music, sounds played backwards, and garbled noises all work to startle the viewer. Use your imagination and then take it down or up a notch.

#226 How to Scare the Viewer Without Showing Anything

Low-budget horror videos don't show you the ever-present evil because it's cheaper. This is the sleazy way out, but the trend is not to show you anything because that builds the expense.

The best approach is just to show a glance of what the horror may be. Instead of displaying the entire being and having the audience say, "Is that all?" just show a portion: a glimpse of a tentacle as it slithers by, a bony appendage as it grapples the wall, or a mouth full of snarling teeth as it salivates waiting for its next victim.

The movie *Alien* did a great job of hiding what was after the crew on the freighter. With a mixture of darkness and the creature hiding in shadows, we never really saw it until the end. By showing glimpses and pieces, we learned that the alien was big, black, frightening, hungry, and downright mean.

Earlier in this chapter we talked about throwing open a door and the viewer being scared in anticipation. But if two kids are peering into an abandoned house through dirty windows and we see a piece of the demon race by, we know the kids are doomed. The situation has been set up, we saw a portion of the evil, and we put two and two together.

In this particular video, "The House," we never even made a creature. Instead we draped a piece of tattered black cloth on a wheeled light stand and moved it past the camera. The viewer saw very tall, tattered clothing, and accelerated movement; that was enough. People would disappear and the audience assumed it was the demon.

The low-budget hit *Jeepers Creepers* used this dread of not seeing to great success. The young heroes see a black "thing" dumping bodies down a storm sewer. After it leaves, they peer into the darkness and the young man falls into the pit.

He hears noises in the total blackness and we are terrified at what he might see. For a long time we don't see anything, just the darkness. When we finally are enlightened (by a flashlight), we see his new surroundings and are appalled. I won't spoil it for you, but this is an excellent example of being scared by what you don't see. You can also be scared by what you do see, but I'll leave that for a sequel.

#227 How to Use Point of View Effectively in Horror

Nothing sends shivers down the spine of your viewer more than subjective or point-of-view shots. Whether we are seeing through the eyes of the hero or another character, it gives us an uneasy feeling when we are detached from our perspective. The subjective shot takes us from our safe, third-person point of view and puts us in the middle of the action (which is where we don't want to be, but should be in horror).

The film *Halloween* made great use of a device called the Steadicam. This gyro-controlled camera could float along, go up and down stairs, and move around corners without the need for dolly tracks, providing a smooth image onscreen. It has been said that one could ride a horse at a full gallop and the Steadicam image would remain smooth, absorbing the bounce.

The use of subjective shots from Michael Myers's (the killer's) point of view brought the viewer into the story as we saw the world through his distorted eyes. The slight wobble that the Steadicam created added to this floating uneasiness. To make the effect even better, John Carpenter chose to put Michael's mask over the lens of the camera so we were seeing through the cutout eyes. We were now seeing through the eyes of the killer when we had no desire to.

In any horror video, we need to see what the evil character or monster sees. Sometimes this point of view is distorted; the character can only see in black and white, in strange colors or patterns, or with more clarity than we can see. By using a subjective shot and displaying what "it" sees, we learn its strengths and weaknesses. Sometimes an alien can see our heart beating in our chest via X-ray vision, or in *Predator*, it sees things in rippled, mosaic patterns. In *Reign of Fire*, the dragon sees images, 'solarized' and in wide, elongated vision, like slanted eyes might. By allowing the viewer to see what it sees, we are adding to the shock value.

#228 How to Use Makeup in Horror

If what you see onscreen isn't scary enough, use makeup to make the frightening even more so. Whether your objective is to show how the "thing" looks, to show blood and gore effects easily, or to show a transformation process, makeup can help make it believable.

Blood is one of the most used (and abused) things in horror. Everybody bleeds when bitten, stabbed, manhandled, or poked, and the more the viewer sees, the uneasier they become. You learned to make stage blood in Solution #206. But if you're doing a black and white video or depicting a creature's blood, what else can you use?

I've used Hershey's Chocolate Syrup to simulate blood because it has the right consistency, is a dark color, and cleans up easily with a spoon and a hearty appetite. Figure 27-3 shows syrup on a wound. This stuff even clots like real blood. By using food coloring, you can change the hue of the blood to suit your desires.

The dead are always depicted as white, gray, or pale blue in complexion, and the closer the camera gets, the more complex the makeup must be. Masks work at a distance but won't be acceptable up close. When wearing a mask, apply black makeup around the actor's eyes to make his or her skin blend into the mask. The actors that played Batman in the movies wore black makeup around their eyes, whereas this was not the case in the '60s TV show. I guess we have matured.

Figure 27-3

Chocolate blood in black and white

This genre is one instance where makeup shouldn't look like it's being worn. People know that these entities don't exist, but we still want them to look believable. A vampire needs a red cape and should have long, pointed teeth for you to accept him or her as a vampire. A werewolf must have hair over every area of skin, a mummy have lots of bandages and rotting flesh, and a zombie have white, expressionless eyes.

All makeup needs to look convincing when seen on camera. What looks great in fluorescent light will look much different under tungsten. This is why Hollywood always does makeup tests under movie lights because it gives the makeup a unique look (which could be better or worse than you have envisioned). Shoot some video footage under identical lighting situations to determine how your makeup looks.

Don't expect your talent to wear anything that isn't safe for use on skin. Some oozing liquids look great on camera but are caustic to skin. The late Lon Chaney, Sr. used to put egg whites under his eyelids to simulate "white eyes." He would strap 60-pound loads to his back for a hunch, painfully twist and pull his skin to play the *Phantom of the Opera*, and numerous other niceties that would never be accepted today. If you won't apply something to yourself, another shouldn't be expected to do it.

In order to get an accurate representation of someone's face—a mask or negative impression of their face may be made.

You can make a detailed mask of someone's face by spreading Plaster of Paris over the skin and allowing it to dry. The mold is then filled with liquid latex and you have a mask that is form fitting to your actor. This needs to be done safely and books on this subject delve into the nuances of breathing and other necessities.

Comb the department stores the day after Halloween and purchase all the makeup you can at reduced prices. Wigs, clothing, scary appendages, and a multitude of colored makeup sticks will increase your arsenal and better prepare you for your next video.

Chapter 28

Comedic Videos

Solutions in this Chapter:

#229 How to Write Comedy

Laughter is the best medicine and people love to laugh. Writing comedy is difficult because what you deem as funny may not be to someone else. My sordid approach to the medium is to try to write in a slightly humorous way, so if one joke falls apart, the entire project doesn't hinge in the balance.

Throughout the course of this book, I've tried to lace my stories and anecdotes with humor. Being subtle and not calling attention to the humor is the best way to write. If you jump up and say, "Look at me—I'm funny," no one laughs. But if you can poke fun at yourself or others without being insulting or degrading (other forms of humor), at least you're on the right track.

When writing a script for your video, look at the entire story and try to find the funny or humorous elements. What can be punched up to be made funny? If something happens to the character, what can you add to it to make it funnier? Kids love repetition, so use that to your advantage. Have your character do something over and over again and still not get it right. Each time she tries something, it doesn't quite work out. We are poking fun at this person in the situation, not necessarily how bumbling or incompetent she is. Once again, insulting humor is degrading and should not be used (unless you make fun of yourself, and only you can sue you).

What can you downplay to get laughs? If people act nonchalant and just go about their business while doing funny things, you'll receive chuckles. Just have them say, "This is no big deal" as everything tumbles down around them. If they realize they are in a funny situation and think it's funny themselves, the humor has disappeared. The old saying is, "No one ever laughs at their own jokes."

How about a person in an awkward situation? You've seen this before a thousand times (every Jerry Lewis movie). The viewer laughs at the character being in way over his head. We've been in situations like this and we know what he is going through. The more he tries to fix the problem, the greater it becomes. It's like he's digging himself a hole he cannot possibly get out of.

Have you tried a situation where someone is pretending to be someone else? In comedy, men dress up like women to get laughs. In the movie *Fuzz,* Burt Reynolds played a cop disguised as a nun, complete with a mustache. Everyone knew he wasn't really a woman, but they played along because it was amusing.

When you begin writing comedy, try to take situations from your own life and look for the humor in them. Even people who never laugh have had humorous things happen to them – use it. What about an awkward meeting, big height or age differences between two people, or someone who tries to help too much? Each of these things has happened to you and you've got to get it on video.

Bounce your idea off other people. What do they think? Maybe they can add to it and make it funnier. When comedy writers get together to come up with a concept, each one builds upon another's ideas until the script is amusing. The original joke may have been a bomb, but by the time everyone puts their two cents in, you have a winner. I wrote a script for the ABC comedy series "Wings." An agent thought it was the funniest script he had ever seen. When the producers of the series saw it, it wasn't funny enough for them. Everyone has their own idea of "funny."

Look at the successful work of others whose style you admire. There have been many great comedy geniuses in the past. Check out what worked for them (without stealing).

The last bit of comedy writing advice is if you don't think something is funny, it's going to be almost impossible for you to get others to find humor in it. But if you believe in yourself (and the gag) you're working on, act it out and see how you can make it better.

#230 How to Use Sight Gags Effectively

A sight gag is something that is visually humorous. It can be an awkward placement of people and things or almost anything that makes you chuckle without having to resort to words.

In most of the commercials I've done, I try to use sight gags to get the joke across. In a 30-second time frame, you don't always have time to explain every joke; a visual that holds up on its own takes up less space than trying to set it up with audio. I could try to explain a sight gag with words, but it loses some of the punch when you try to spell it out. It's like telling a great joke to someone, and they give you a blank stare, not understanding why it's funny. Once you explain it, the joke loses its humor.

Figure 28-1

A sight gag

Take a look at Figure 28-1. Anything I would write explaining the picture would take away from the sight gag. I used this image in a commercial and I know you could come up with 61 possible captions or one-liners to use for the ad. Just like Gary Larson in his book *The Far Side Gallery*, he had a silly cartoon image with three possible captions beneath it, each one being extremely funny. How do you choose the right one? Which one works best?

The next image, Figure 28-2, is another example of a sight gag. You are seeing something that is out of the ordinary. Not all of my sight gags involve toilets, but sometimes my humor is very flushing.

Anything out of the ordinary, out of context, or visually strange may be used as a sight gag. The important thing about these gags is the less said, the better. You want the visual to carry all the weight for its shock value. These gags are always great endings because they leave the viewer laughing. What better way to end the day? Maybe that's why my picture isn't on the back cover of this book (either that or I just don't show up on film).

#231 How to Edit a Comedy for Pacing

A comedy can fall apart if too much time is spent on one shot, if it is cut too quickly before the viewer understands the joke, or if the pacing is erratic.

Figure 28-2

Another sight gag. What's wrong with this picture?

Consistency is still the name of the game, and you control all the strings with editing.

In order to establish pacing, storyboard your scene after the script has been completed. Suppose you have a scene of a valet working in a parking lot. First, establish the shot so the audience knows the locale of the scene. In our case, a beautiful new BMW Z4 pulls up and the driver tosses the keys to our talent, the valet. This all occurs in a long shot, up to the point where the valet catches the keys.

We cut to a close-up as the keys land in the valet's hands. The close-up identifies the young, eager valet. He smiles as he grabs the keys and we can tell from his expression that he is looking forward to driving this exotic car.

Cut back to a medium shot as he gets in the car and can't quite fit behind the wheel. In a close-up, he fumbles to find the seat switch to make himself more comfortable. Hitting every available switch, he operates the alarm, headlights, wipers, convertible top, CD player, and so on. The valet being out of his element in a car this expensive is the beginning of the joke. We don't want to dwell too long on this shot, just long enough for people to understand he is confused.

We need to get a reaction shot of the owner as he stops in his tracks and slowly turns around to see where all the noise is coming from. Even if this driver is a known star, don't spend too much time on his reaction. Instead, cut briefly to him in close-up so the viewer can see the terror on his face.

Cutting back to the valet in the car, he notices the owner racing over to take the keys away. He must quickly get everything in control before he loses his job. Just as the owner gets to the car, the valet throws it in first (we see this in a close-up) and punches the gas pedal (also shown in a close-up).

Back to a long shot, we expect to see the car peel out and move forward. Instead, the car moves backwards, much to the shock of the valet and the owner (both shown in their own close-ups). Leave just enough time on the valet's face for him to express, "I thought I was going forward! I put it in first!" Then cut to the owner's face, him thinking, "Oh my God! What is he doing?"

The car rolls backward out of frame; we immediately cut to a shot from the valet's point of view through the windshield, trying to find a way to stop the car. Then we show a close-up of the owner's face as the screeching sound of brakes, splintering glass, and crunching metal is heard. He closes his eyes as he witnesses the demolition of his pride and joy. We never see the mangled car; we just hear the noise (that saves a lot of money). The viewer imagines something worse than we could show.

The camera pulls out to reveal the valet walking up to the owner, now in shock, and sheepishly dropping the keys in his hand. The owner doesn't move. He just stands there staring at the keys. The camera stays on this scene for a moment (pacing) and the valet walks back into frame and speaks to the comatose owner. "Ah, sir! You may want to get that transmission checked. I think something's wrong with it!"

Still speechless, the owner stares at the keys in his hand and then up at the valet as he walks away. The owner never says anything, but we feel the pain he is in from what has just occurred. By establishing a pace through cutting, we tell the story humorously without having to show anything too costly to create, using very little dialogue, and some sight gags to get the point across.

Chapter 29

Period Videos

#232 How to Create a Specific Time Period's Look

Because we're never happy with what we currently have, we always want to change it. When shooting a period video, you must make the audience believe they are watching something occurring in that particular time. How do you make this believable?

No matter what the particular time period is, it must be convincing. The props, backdrop, costumes, and actors must look the part. If anything on your set stands out of place or looks like something recent, you've destroyed the illusion. Let's look at a particular period piece and see how it was made.

My master's thesis project was a film called *The Butler Did It*. This late-1920s comedy centered around a young man trying to get the nerve to ask his girlfriend to marry him. The film was done using high-speed lenses and low-light cinematography, and I thought if I could present a short film utilizing these elements, the world would be a happier place.

The first step in any period video is research. Learn all you can about the particular era you want to depict. What did people look like (clothing and hairstyles)? How did they act (customs, where they went for entertainment, and how they socialized)? What kind of music (if any) did they listen to? What kinds of items would be on the set or location (props and vehicles), and how did they talk (particular language, slang, or sayings)? Once these questions are answered, you'll have a better idea of how to create your time period. Remember, it must be believable in every way.

This research can be done from old books, movies from that period, the Internet, and interviewing people who lived during that time. If you are doing a video on the Roman Empire, it might be difficult to find people to interview, but I think you get the idea.

When writing the script for *The Butler Did It*, my professors thought it should be in black and white because the films of that the period weren't in color. But since people actually saw in color in the 1920s, I wanted to depict that. They also suggested I use title slates for dialogue, much like you would see in a silent film. Again, I wanted it to be more realistic and the only reason films of that period had no audio is because it wasn't invented until 1927. I was a rebel, what can I say.

However, I wanted all camera movement to mimic that of the period. This was a time before zoom lenses, Steadicams, complicated boom shots, and so on. Instead of zooming into a shot, I would dolly in the same way they did. The lighting would also be from the 1920s with practical light being the main light source. The globe table lamps I used gave off a warm, pleasing light, so I gelled everything with bastard and amber gels to make life warmer.

The set, or in my case location, had to look like the period. I was able to obtain a mansion (I had to give it back) from the early 1920s whose 94-year-old owner vacationed in Florida during the winter months. The walls were solid walnut, the opulent furnishings were from the same period, and everything about this place was 1920s. I really lucked out.

It turned out that the gardener on the property owned an original 1928 Model A truck that I was able to use in the film. The vehicle was exactly as it would have looked in 1928 with the exception of a current state inspection and license plate. I didn't want to show that this vehicle was a truck rather than a car, so I just shot it from the front and side and framed out the tailgate area.

The last step was getting period clothing, which is discussed in Solution #235. Once the actors had been cast and the set was lit, we began shooting. I started with the exterior scenes because it was winter and I had a short period of daylight. Obviously, in a period video, frame out anything that looks like it's from the twenty-first century. I shot subjective shots through the Model A's windshield as it traveled down dirt roads, keeping an eye out for planes, modern vehicles, or anything that didn't look 1920s.

One important lesson I learned was do everything in your power to preserve the props. Our actor, having a difficult time exiting the truck, stepped on the running boards over 20 times until the shot was convincing. He also had tiny particles of ice and gravel on his soles that he grounded into the original paint on the running boards, which had survived 70 years before we arrived. Since this car cost more than my wife (who I didn't have yet), I had to make these scratches disappear. With today's polymer polishing compounds, I restored the running boards to their original condition with hours of buffing. Figure 29-1 illustrates the car that caused all the trouble.

Figure 29-1

Watch out for the running boards!

Everything else about the program went off without a hitch. The sets were real, the dialogue sounded right with the correct vernacular, and the warm glow from the lights made me think I was in that time period. Figures 29-2 and 29-3 show scenes from the interior set.

Looking through the viewfinder, I believed I had achieved the look of this time period. With modern equipment and technology, you can achieve a period look for your videos with less trouble than you might believe.

#233 How to Blend a Modern Backdrop into a Period Video

How do you change something that's current and make it look like it belongs in the past? The easiest way is to take everything out of the set that doesn't belong. It sounds easy, but it's a little more difficult than that.

This is where continuity is extremely important. Someone has to be constantly watching for things that don't belong in that time period because the

Figure 29-2

Getting ready in the bedroom

viewer will notice. In an old *Gunsmoke* episode from the 1960s, a fight scene occurs on a dirt street. As the actors hit the ground and throw dirt into each other's eyes, we notice the feet of the crowd of people around them. This is normal and people will gather when they see a fight; they probably did the same thing in the old West. However, one of the hired extras had visited wardrobe and received a period dress, but she neglected to change her 1967 vintage black pumps. Most women of this era thought a pump was something you got water from, not a shoe. Immediately, your eye goes directly to this woman's feet. My point here: Look for tiny things that don't belong.

In the movie *Somewhere in Time*, Christopher Reeve just has to look at something from the nineteenth century and dwell on it to be transported to that time period. As long as he believes he is there, he will remain there. He takes great precautions to stay with the woman he loves (Jane Seymour). He goes to antique shops to get money, clothing, and jewelry from that time period; everything is as it should be. That is until he reaches into his pocket

Figure 29-3

The dining room scene

and finds a penny from the 1970s. This remembrance of his current time takes him back to the present. Don't let this happen to you!

A set can easily be changed to depict the correct era; the outdoors is a little more challenging. If doing a medieval sword fight in the woods, a passing plane, a highway nearby, or apartment complexes beyond the ridge must be removed (from the shot, not permanently). Watch for telephone, cable TV, or electric lines being in your shot. The same applies to cell towers, satellite dishes, or anything that will cheapen your period look.

Comb the entire area just looking for things that don't belong on the set or location, including costumes or hairstyling. Mistakes like this turn up in films all the time and people laugh at the inconsistency. Don't be a statistic (I sound like a public service announcement); check and recheck everything. After doing your research, you'll know exactly how the chariot race should look. Just keep the kid on his Big Wheel out of the background.

#234 How to Use Black and White Video Effectively

Black and white images always seem to conjure up a special look. Whether is a film noir scene from the 1940s, an old comedy from the 1920s, a dream sequence, or just a depiction of a simpler time, this monochromic look is unique.

We can easily achieve this same look with video. By removing the chroma, you essentially have black and white images or those that lack color. I recently used these varying shades of gray in a corporate video.

Wanting to create a 1940s detective set, everyone agreed black and white video was the way to go. A junior executive was being hounded by his boss because he didn't look at problems analytically. As the camera pulled away from him sitting on the edge of his desk, all the color disappeared from the shot. This gradual dissolve from color to black and white occurred in editing.

The actor finds himself on a private detective's desk in the 1940s. The PI, who turns out to be his boss, has to explain to him how a detective would solve his corporate problems. She is dressed in period clothing and wearing makeup that works for black and white video.

Her lipstick, normally red in color video, was black for our black and white segment. If shooting in black and white, you should accentuate the makeup, props, and clothing to look realistic in the "shades of gray" world. The walls were painted gray instead of light yellow or green, the clothing was drab and colorless, and the makeup on his boss (lipstick and eye makeup) was darker than normal.

Because we wanted him to stand out in the black and white world of the 1940s, the color of his clothes was more vibrant (although still just shades of gray in black and white). We wanted to create something different about him because of his clothing, but in a subtle way. We could have left him in color and had everything else in black and white like in the movie *Pleasantville* (selective colorization), but since he was part of this new world, he would look the part, only slightly different.

The *Superman* TV series from the 1950s with George Reeves was shot in black and white its first season (as was *Gilligan's Island*, *The Beverly Hillbillies*, and *The Andy Griffith Show*) and the rest of its run was in color. To make his blue and red costume look peachy in black and white, they decided to make the red cape and S emblem dark brown in color and the rest of the normally blue fabric light gray. If you see these episodes on TV Land, you notice how dark and strange his costume looks in black and white.

In the black and white thriller *The Incredible Shrinking Man*, the tiny hero impales a tarantula with a needle. Lying on his back underneath the spider, he sticks the needle in and spider blood pours all over him. Since this was a 1950s low-budget thriller, chocolate syrup was used for blood. It looks and flows like the real thing (unless spiders do have chocolate blood), and no one can tell the difference.

Experiment with how different things look in black and white, and make your video that much better by choosing colors that look better in black and white.

#235 How to Use Costumes in Period Videos

Costumes are the best way to portray a period video. If you put someone wearing a gladiator costume in the middle of a shopping mall, people will believe (without an verbal explanation) that this fellow was transported from his time period to ours. To make this even move convincing, have the actor look misplaced or confused in his new surroundings.

People expect the costumes to suit the period. In fact, just by looking at someone in costume in an empty studio, we can tell what time period they are from. Dress someone in animal skins, we think cave people; suits of armor, the middle ages; stove pipe hats and flowing gowns, the nineteenth century. In the twentieth century, every decade had its own look, from the 1920s to the 1990s. This is the easiest portion of a period video.

For *The Butler Did It*, I found period dresses and suits from the late 1920s at estate sales. The hairstyles also had to match. My actress's long, beautiful, red hair was tucked under a short, blonde wig because women in that period did not have long hair, especially flappers. Long hair was considered a sign of a "loose woman." The male actors had short hair combed back and slicked down, with no facial hair other than tiny, pencil-thin moustaches. In keeping with the style of the era, I wanted everything to be credible.

I believe it's important to have someone responsible just for costumes on any period video. You cannot skimp in this area, yet you might have too many other things to deal with and not be able to devote the time that costumes require. Contact high school theatre groups, colleges, or costume shops to find what best fills your needs. Just like Hollywood, do a costume and makeup screen test to see if your actors are authentic and the costume fits them correctly.

Sometimes a slight tweak to the hairstyle or makeup will make a costume look better on someone. Once again, do research to determine the best costume for your characters.

Solutions in this Chapter:

Chapter 30

Romance Videos

#236 How to Use the Close-up

No better way exists for helping the viewer identify with the character onscreen than the close-up. The next solution (#237) will discuss other ways of identification.

The close-up almost gets us into the mind of the character because we can see their expressions, facial features, and eyes to determine how they're feeling as well as what they may be thinking. You really can't use a close-up too often. Once you establish a shot so people know where you are, move into a close-up. Entire films have been shot entirely in close-ups and it makes the audience uneasy, but we do want to move in close and see what's happening.

To use a current example of an effective close-up, let's look at Christopher Nolan's film *Insomnia* with Al Pacino and Robin Williams. In one scene, Al Pacino has to call his dead partner's wife to tell her he was killed. He doesn't want to make the phone call, he's uncomfortable, and he has to do it in a police station filled with people. Obviously, the best way to do this shot is in a close-up. We want to see his brow sweating, him fumbling with the phone cord, and other little nuances showing his frustration.

Nolan chose to shoot this close-up slightly differently. Instead of having the camera across the room (where it was for the wide shot) and using a telephoto lens (200mm), he decided to bring the camera two feet away from Pacino and use a normal lens (75mm). The shot now becomes more intimate and claustrophobic because the camera is so close to the actor, much like someone invading your "personal space." We never like someone to talk to us three inches from our face; we want to back away, but Pacino can't in this scene, just like he can't back away from making this phone call. As long as his face fills the screen, the shot is still considered a close-up. But by moving the camera in tighter to the actor, he feels more uncomfortable and his acting method changes slightly to a better, more uneasy performance. The director made a great shot even better by moving the camera in closer instead of staying at a distance and flattening the image with a telephoto lens.

A close-up makes us feel what the character is going through. If he is looking at something, the viewer wants to see what he is looking at in a close-up. Like the film term *juxtaposition of image,* whenever you cut two shots together, the viewer will assume they are related. If you show the backside of a donkey, and then a close-up of your Uncle Sidney, people will assume he's a donkey butt (I did that as nicely as I could).

Don't be afraid to use the close-up. Remember, you control the shots.

#237 How to Make the Viewer Identify with Your Characters

In the previous solution (#236), I mentioned how important the close-up is in character identification. Two other methods are available for making the audience feel they are in your character's shoes: the eyes and the subjective shot.

Since you know all about close-ups, getting even closer and showing the talent's eyes will pull more weight in the character identification game. The eyes are the windows to the soul and we can tell what someone is thinking, feeling, and expressing just by looking into their eyes.

In a love scene when a couple is breaking up, we see a close-up of the woman's eyes. They are full of tears and we know she is saddened by what is transpiring. The females in the audience will identify with her because they see her eyes, the pain and suffering, and may remember when they went through a similar situation. The males, looking at the man's eyes, may recall when they did the same thing. A man may seldom admit he is looking into another man's eyes, but when shown in a close-up, there's nowhere else to look.

The same thing applies in any James Bond movie. The men all identify with 007 because they see he's cool, calm, and always in control; they want to be like him. Women equate themselves with Bond women who are attractive, larger than life, and can do anything.

The goal is to get the men in the audience to relate to the male characters and the women to the females. By showing these characters in various situations and what happens to them (through their eyes), we have total identification. The viewer begins to think about what he or she would do in a similar situation.

The subjective shot does exactly the same thing. Instead of seeing their eyes in a close-up, we are seeing life *through their eyes* in a point-of-view shot. We have now traded places with them on the screen and can do anything they can do. Like Bond, men can expertly ski, fight off villains, and woo the women. The women are independent and sexy, and can handle themselves as well as, if not better than, a man.

Once again, the entire video can't just be seeing people's eyes, because we need the rest of the body also. A project shot totally in point of view (like *Lady in the Lake* in the 1940s) becomes trite because we need to see the outside of the body once in a while. A variation of both will make a winning video that will have the viewers believing they are playing the parts too.

Chapter 31

Lingerie, Nudity, and Sex

#238 How to Shoot Lingerie Without Lingering

I used to feel misplaced and uncomfortable when shooting a lingerie ad or commercial. The models would come in, dressed in their lingerie of choice, I'd shoot them, and we'd all go home (not together).

The whole idea is to have the garments (lingerie) look as appealing as possible without drawing too much attention to the model. This needs to be achieved by lighting, color temperature, skin conditioning, and atmosphere. Figure 31-1 shows how we successfully lit our model and you pay absolutely no attention to her.

Make the set as comfortable as possible for the talent and try to focus on your skills in video rather than how uncomfortable you may be in this situation.

#239 How to Light Underwear

Making a model dressed in skimpy clothing look appealing doesn't take too much, but lighting is key (another light joke). Warming her skin tone makes a more inviting (she may not be), friendlier (she may not be), and eye-popping model (she is). This warmth can be done by gelling the lights with straw, orange, or, my favorite, bastard amber gel. These are subtle warming changes. If you color balance your camera to daylight and use tungsten lights, your set will be too warm and orange. The idea is to warm up the shot a bit, not bathe her in orange light.

The model's skin will naturally absorb less light than her lingerie (unless it's white). Figure 31-2 and 31-3 show the same model dressed in white and black. Notice how identical lighting makes her skin appear totally different because of the clothing.

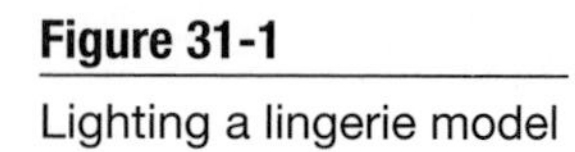

Figure 31-1

Lighting a lingerie model

#239

Figure 31-2

Joy dressed in white

Figure 31-3

The same Joy dressed in black

Black absorbs more light and appears darker, whereas the white lingerie fluoresces more.

Black and white behave differently under light, so flag off the light so it doesn't hit the skin quite as much when shooting white lingerie because the skin will overexpose (unless the model has very dark skin). Don't use flags with black undergarments because all the needed light will be absorbed and the skin will still look natural with light falling on it.

#240 How to Control Color Temperature

The color temperature of your set helps set the mood. When I shoot white lingerie, I balance more for daylight and cool the colors more toward the higher (blue) end of the spectrum. Don't balance for tungsten and use daylight illumination; that's far too blue. Instead use Booster Blue gel on tungsten lights and still balance tungsten.

Black or dark clothing requires just the opposite: a warmer approach. Using the gels I mentioned earlier (such as bastard amber) will lower the color temperature to the orange end. This also works if your model has dark or extremely pale skin (more on lighting skin in Solution #241).

#241 How to Give the Skin Texture with Light and Oil

The model's skin should glisten and any imperfections should be hidden with makeup. I shot a klutzy lingerie model that had the habit of walking into things because she refused to wear her glasses. Each impact created a welt on her skin that your eye immediately noticed. Since you want the viewer to look at the clothing and not the wearer, anything distracting on the model must be disguised or it will take precedence. The same thing applies to freckles, moles, and other birthmarks.

Baby oil is the preferred way to shine the skin rather than water. Water droplets look too much like perspiration (a no-no in lingerie) and are more transparent than oil. Baby oil (squeezed from babies?) clings to the skin and light reflects off its surface. Bodybuilders use baby oil when performing because it accentuates the muscles more (while hair detracts).

#242 How to Give the Background Atmosphere

The set could be smoky, brightly lit, colorful, or almost anything that will draw your attention to the clothes the model is almost wearing. In this situation, contrast is the most effective. If the model is wearing brightly colored lingerie, the background should be slightly darker. If dark apparel is chosen, a saturated color works well.

Look at advertisements in magazines or Sunday supplements. What works and doesn't with the examples you see? The most important thing to remember is to accentuate the lingerie. If the model is unhappy, frowning, or doesn't seem to want to be there, the viewer will notice her rather than the lingerie. As Mr. Rork in *Fantasy Island* always said, "Smiles, everyone, smiles."

The model should also match the lingerie she's modeling. A full-figured woman should not be modeling skimpy clothing, as a petite woman should not be swimming in her garments. Use common sense and you'll know immediately if something isn't working; just ask your wife. Figure 31-4 illustrates all that goes into shooting someone wearing very little.

Figure 31-4

Never has so much been needed to shoot someone wearing so little.

#243 How to Make Your Actors Comfortable

A lingerie shoot is an awkward situation to say the least. How can you make your model feel comfortable with a bunch of clothed men on the set?

I'm assuming everyone involved with the shoot is a professional and is probably getting paid because they are very good at what they were hired to do. Each crew member is focusing on their specific job, but that has to come across to the model.

Have the set ready for the model when she arrives. Fine lighting adjustments may be made once she is on the set, but have the major decoration and lighting ready to go. When the talent changes into her lingerie, have her wear a robe on the set until you are ready to begin shooting; it isn't necessary for her to remove the robe before this time. My models (I like the way that sounds) have the majority of their makeup adjusted while wearing a robe. Fine adjustments will happen when they are just dressed in lingerie. Figure 31-5 shows how natural this looks.

The temperature of the set should be warm and comfortable without any hot or cold extremes. You don't want her to bake or freeze on the shoot because the end viewer will see that. Since the model is wearing very little, adjust the temperature to suit her, not the rest of the crew.

Only the director, makeup person, and camera operator should have direct access to the model. I don't mean that no one else on the set can talk

Figure 31-5

Makeup and lingerie, just like Christmas at my home (just kidding)

to her, but if only one person is giving camera direction or adjusting and tweaking lights, she won't feel as if she is on display. Common sense comes into play here. This shoot will be very pleasant for everyone if decorum is maintained on the set and everyone knows their roles.

When I directed my first lingerie shoot, I spoke to the crew beforehand like I was talking to children. "Don't ask her silly questions." "Don't try to be overly helpful." And basically, "Speak when you're spoken to."

On another lingerie shoot, when I was adjusting lighting on the runway, the model was practicing her walk for the director. As I was tilting a gelled light with a clothespin in my mouth, she slipped on the runway. It's not often that a woman in high heels and underwear falls out of the sky into my arms, but this one did. It was the most awkward moment in my life. I was uncomfortable. What do you say at a time like this? Luckily, my wife told me what I should have said. The next time it happens I will say it.

#244 How to Shoot Sex Scenes Without Sex

Almost every Hollywood film has a sex scene as part of the story. They have become commonplace and the viewer expects them. If your video requires one (as per the script—don't just add it because you want to), we'll discuss how to shoot it realistically.

The *Motion Picture Association of America* (MPAA) has developed a rating system to determine what the audience may be shown in a specific picture. A film rated G is open to anyone (general audiences), a PG rating means parental guidance is suggested (no nudity), and PG-13 allows children under 13 to view the material with parental guidance (extremely brief nudity, the backside, or upper torso). An R rating meaning restricted (under 17 must have a parent or guardian due to brief female frontal nudity or extremely brief male nudity) and NC-17 (formally X, allowing no one under 17 and almost anything goes). Issues with language and violence also come into play with the rating system.

When shooting a love scene (it sounds better than a sex scene), the more that is left to the imagination, the better. It's not needed that you dwell or really show anything, because most adults "know" what is happening anyway. I've shot quite a few sex scenes (before I was married) and under no circumstances should the couple actually be "doing" anything. The nature of video is that these people are "acting."

By showing close-ups of faces, hands, arms, legs, backs, and torsos with a moving, flowing camera, a sex scene is easily simulated. Without showing the whole body or having totally nude actors, small brief glimpses of the body in close-ups usually suffice. I've shot love scenes from as simple as a woman taking a man by the hand and leading him into the bedroom and the door closing, to couples in bathing suits under the sheets, to seminude actors shot through a series of close-ups. Each of these was the lovemaking process without showing sex.

Only pornography shows close-ups of sex acts and I believe there is greater mystery and creativity by not really showing anything. This usually makes your actors more comfortable.

Sometimes showing what occurs up to sex implies that the couple will be engaged in that shortly, but the scene fades out and we are left to believe or assume what is going to happen next. Or, on the other hand, if the male or female moves away from their partner in a bed after an embrace, we imagine they have just finished. Through the magic of video, just showing a portion or implying what is happening is usually enough.

#245 How to Shoot Nudity

This is a very touchy subject that most people don't want to deal with. What do you do if your script requires nudity? How do you get talent?

If hiring professionals, their pay rate usually doubles if nudity is involved. Most have done nudity before and you are paying them for the "right" to videotape their bodies. But if your budget has minimal funding, how can you "fake" nudity? Once again, this is easily accomplished by only showing little bits and pieces. Let's look at partial nudity first (no matter how I word it, it still sounds strange).

If your scene involves topless nudity, I've often shot the scene from her back. When the audience sees a bare back (not on a horse), they assume the front is also unclothed. Why show it? In this case, the actor can cover up her breasts by wearing something that cannot be seen from the back.

In all the shower scenes I've shot, the male or female is wearing a bathing suit. With her top straps down, you can tilt the camera below the shoulders, simulating nudity without displaying it. Even if a portion of the breasts is exposed (like what is usually seen in low-cut clothing) as long as everything isn't visible, we get the picture.

When showing an actor's bare backside, it's a little tougher. When viewed from behind, it's obvious that the front is naked also. By keeping the talent facing away from you, we never glimpse the front. Many times studios require this kind of nudity to keep a PG-13 rather than an R rating. Nudity is still involved, but it is limited.

If actors are nude as viewed from the rear, how do you make them comfortable and not embarrassed by crewmembers viewing them from the front? In a Michael J. Fox film, he walks to the door naked (as seen from behind) and greets who he thinks is his girlfriend. Instead, it is a young Girl Scout selling cookies. Upon seeing him, she screams and runs off. On the set, the girl did not see him naked.

Hollywood has devised ways to hide things from those who shouldn't see. Mr. Fox wore a cup that was fastened to his chest and legs (which from our vantage point, we couldn't see). It might look strange to the crew seeing him walking around with this covering taped to his body, but it helps those who are modest.

Women have used similar coverings when on the set. It allows nudity without showing anything: the perfect marriage. In a long shot, nudity is shown at a distance. Flesh-colored body stockings are worn because the camera is so far away; all detail is lost. In scenes that involve public nudity, this is usually what is worn: wearing a body stocking without being nude.

The last case involves bodies on a slab in the morgue. These shots also necessitate nudity and a living person portraying someone who is deceased. Usually shot from below, we can see the person is naked, but the camera is far enough away that features are blurred. The other approach is just to show pieces (arms, torso, legs, and so on). By seeing the establishing shot of the nude person, they can wear clothing for the close-ups because only parts are shown.

Various other tricks can be used to show nudity while hiding it. Use your imagination and come up with new ways and I will include them in the next edition of this book (or maybe not). Figure 31-6 shows a still from that shower scene.

Figure 31-6

Naked but really clothed

#246 How to Hire a Crew for a Nude Shoot

If you have no way around it, how do you hire a crew for a nude shoot? When the line gets short enough from the volunteers, choose only those that need to be on the set for that particular scene.

I was shooting a scene that involved a man, a woman, and an ice cube (not the singer). Nudity was involved and the director wisely decided that the set should be closed. No one would be there who wasn't vital to getting the scene shot. That meant both actors, the ice cube, the director, and me as the camera operator.

Once the director of photography lit the scene, she didn't have to be there because only the camera operator was needed. The director had to coach the action and the ice cube; it had a mind of its own. The actors would obviously be self-conscious if several crewmembers were gawking during the intimate scene.

To get the lights, boom microphone, and props in the right places (the ice cube again), rehearse the scene with the actors clothed. Anything can be adjusted and tweaked with lighting, props, camera, and sound without making the talent uncomfortable by being naked. When the director is ready to roll tape, that's when the unessential crew disappears. In our case, the audio was dubbed because no dialogue was spoken.

In another strange situation, the actor was very uncomfortable performing the scene. The director coached her with clothing on (they both had clothing on, the director and the actor) until he got the performance he was looking for. With 12 of us in the bedroom for the scene, if the director took the time to clear the room, have the actor disrobe, and start the camera, the spontaneity would be lost, or so he believed. Instead he whispered to the actor to try the scene one more time clothed and if it worked, he would nod his head and she would disrobe quickly under the covers and they would shoot the scene for real. The only direction I got from the director was to "keep the camera rolling until I say cut."

She did the scene perfectly, he nodded his head, and she undressed in 3.2 seconds, tossing her bathing suit in the corner as I continued shooting. No one left the room, I kept rolling, and the scene went flawlessly. The key? Great direction, great acting, and a darn good camera operator who follows directions.

Chapter 32

Audio Challenges

Solutions in this Chapter:

#247 How to Hide a Microphone

If you're anything like me, you don't want the viewer to see the microphone. Everyone knows a microphone is used to record audio, but it should be heard and not seen. Half the fun is trying to find new and innovative places to hide the tiny pickup device.

On a recent shoot, I had the opportunity to assist on the sound portion of a Digital Betacam spot, and I placed the microphone on the talent. The sound source was achieved by miking the talent with a lavaliere mike. A lavaliere has an omni-directional pickup pattern that is acceptable in the close proximity in which it is used. But how would we hide the fact that she is wearing a microphone? I guess it would look rather tacky if she waltzed down the stairs with this big mike tacked to her blouse. Unlike the lavaliere microphones of the 1970s, the newer units are smaller and more powerful. But you still never saw Mrs. Cleaver, Donna Reed, Batgirl, or Mrs. Brady wearing a microphone, so we had to hide ours.

My first choice was to plant, I mean place, the mike in her hair. The talent's lovely stock of hair would easily cover the mike cable, but when I asked for the staple gun to secure it, she backed away. Broadway actors always use hair mikes because it allows movement and the object recording the sound remains hidden. I don't always agree with the use of hair mikes because the pickup unit is behind and above the talent's mouth. If anything in the talent's hair makes any noise, you hear that too. I prefer placing the mike closer to the source of the sound.

Since a hair mike was out, I lightly gaffer-taped the tiny instrument under the left side of her collar. Since she would be turning her head and speaking to the left, this proved to be the best side. Every actor has a best side; I guess this was hers. The mike would easily pick up the sound emanating from her mouth and would still be concealed from view. I wrapped the mike with folded-over (looped) gaffer tape so the mike itself was cushioned against the tape and not the inside of her collar. If I had taped it directly to the collar, any movement might cause a rustle, and since no cattle were involved, there would be no rustling on my shift.

If we had used a conventional lavaliere mike, the XLR cable would have protruded from her collar and could easily be seen. Instead, a wireless unit was used. The thin black cable from the mike to the sender unit snaked from under her collar, up over the middle of the collar behind her head, and down her back, and the nine-volt power unit clipped to her waist. It's best to warn the talent before you slip the cold unit down his or her back. Being thoughtful, I warmed ours on the stove before it made its descent.

With the mike and all the cabling safely hidden, we were ready to start recording some sound. When recording a digital signal, it's imperative not to override the gain; this just causes distortion and clipping. If you usually keep your VU meter sound around 0 in analog, back it off in digital to -3 or -5. In Figure 32-1, can you spot the hidden microphone?

(Continued)

#258
How to Record Dialogue

#259
How to Record Ambience

#260
How to Record Music

#261
How to Record Sound Effects

Figure 32-1

The microphone is cleverly hidden, or is it?

#248 How to Eliminate an Echo

Recording audio in tombs, vaults, or large rooms usually means the audio will sound rather hollow and full of echoes. How do you make the cavernous sound more appealing?

I had a shoot in a million-dollar home (the owners knew I was there) where the surroundings were fit for a king. All the woodwork from the hardwood floor, 15-inch diameter columns, and open, trussed vaulted ceilings was solid cherry. Costing more than the GNP of most small countries, this home served as the location for an older woman's testimonial on how much she liked the client's product.

Although she was miked with a lavaliere, her voice reverberated around the catacombs and sounded horrible. When shooting a close-up, we could have used a boom microphone, but any movement from the crew caused the expensive floor to squeak. The narrow pattern of the shotgun would have recorded that extraneous noise. Since hard surfaces (wood) reflect and bounce sound waves, causing them to echo, we needed something that would absorb sound and virtually deaden the room.

We grabbed every packing blanket we could find and covered the floor and walls, surrounding the talent with a padded environment. By gaffer-taping the heavy blankets to the wooden surfaces, we deadened a very "live" room. Although it looked strange off camera, the blankets created a much more pleasing sound to the room. Figure 32-2 shows how strange all the blankets looked in the room, but if you hold the picture to your ear, you will hear nothing (see, the blankets worked!).

Figure 32-2

Sound blankets keep the room dead without hurting the actors.

#249 How to Use the Left or Right Speaker More Effectively

You can enhance any video production by having the sound that appears on the right being placed in the right speaker. Most everything is recorded in stereo, so keep that left-right channel unmixed until the bitter end. By doing this, you have more control over the sound, and that gives you options in editing.

With NLE systems having 99 audio tracks, why cram everything on just 2? Without making the audio muddy, separate every effect on a track of its own. In the old days of film editing, every person and each effect was on its own track. In editing, this gives you leeway you never thought possible. It may take a little more time in the edit, but this also separates the amateur from the pro.

Play with where you want the sound to come out (which speaker—left or right). If the person on the left is speaking, pan your audio over to the left track. Even if you recorded in mono, all *nonlinear editing* (NLE) systems give you the option of track separation/movement. If you can have an effect move from one speaker to another, do it. A little audio sweetening can make a production come alive.

And speaking of sweetening, be careful on the Foley effects. A lot of spots overuse and overmodulate a created effect onscreen. Use common sense; if an effect should be amplified or recreated, just do it at the level it needs to get noticed. An effect that overly calls attention to itself is an effect that didn't work.

Look at other people's work (don't copy it; look at it). How did they achieve a certain effect? What worked and what didn't? I don't think you'll ever run into a situation where you are using too much sound on a video. Every effect has its place when used correctly.

#250 How to Mike Ventriloquists

You may not have lots of opportunities to videotape ventriloquists, but if you shoot even one, reading this may be time well spent.

This particular cast involved a dummy. This wasn't one of my relatives, but a ventriloquist and his wooden, inanimate accomplice. This ventriloquist was hired to entertain another gathering of people (corporate people will stop at nothing to entertain others). This company goes out of its way to hire strange characters to be presenters, and it's always my job to shoot or capture the sound.

When setting up the mikes in rehearsal, my assistant attached a lavaliere to the ventriloquist and another one on the dummy's lapel. He expected the ventriloquist's voice to actually come out of the dummy. From my younger days, I learned that ventriloquism is actually an illusion and a voice cannot be "thrown;" both voices actually come from the ventriloquist. I told my assistant to leave the mike on the dummy, but it would remain inactive (just like the wooden talent). So in this example, make sure you mike where the sound is actually coming from, not where it appears to come from.

To keep the illusion, it's always good to have a spare inactive (batteryless) mike to attach to a dummy. When the dummy speaks (why did I say that?), people will naturally look at him because it's the polite thing to do. Since human beings are made this way, this takes some of the pressure off the ventriloquist. Just make sure before you start which one really is the dummy.

#251 How to Mike Without Using a Microphone

Times will occur (for whatever reason) when you can't easily attach a microphone to the talent. Sometimes even a shotgun microphone on a fishpole isn't practical. What do you do in instances like this? Besides giving up, several ways exist for getting high-quality audio without using the normal methods.

Recently, I had the dubious honor of being a sound person at a corporate meeting. The audience was filled with 300 corporate types who were all there to hear a company "State of the Union Address" as far as it pertained to them and their corporation. A stage had been erected in the front of the auditorium and two four-foot by five-foot video projection screens flanked the stage. The main action would happen on the stage and the three video cameras recording the event would display the images on the projection screens. This type of event had happened many times before, but this time the usual method could not be employed.

Normally the chief financial officer (CFO) of the company stands at the podium and presents the facts and figures as slides or graphs are shown on the projection screens. This setup is easily miked with a microphone on the podium or the presenter wears a lavaliere if he or she chooses to wander or pace the floor. Most times these lavaliere mikes are wireless so the presenter can dance to their heart's content without tripping over XLR cables.

The presenter this year was an alien. Now when I say alien, I mean someone from another planet. An actor donned a silver costume (lighting it for video was a real treat) and wore a 15-pound papier-mâché head. This mask-like head had two small air holes for breathing (not seeing) and not much else.

This alien would arrive in a cloud of smoke and vapor, walk to the center of the stage, and speak to the crowd. Because the mask had no moving parts, the actor couldn't open his fake mouth to speak; all he could do was gesture.

In rehearsal, I tried to put a tiny lavaliere mike under his mask so he could speak, but his voice sounded muffled inside the head. The CFO thought this muffled voice gave away the trick (like someone in the audience actually thought we hired an alien and it wasn't someone in a costume). There really wasn't any way we could mike this thing without having the sound distorted; after all, his voice was coming from behind the mask. To complicate matters, the CFO also wanted the alien's voice to appear "alien," not muffled but sounding echo-like and metallic. Getting the reverb could easily be handled with any mixing board and reverb unit, and the metallic sound would be a little more difficult.

The easiest way to capture the alien's voice was to record it and play it back. Miking the alien didn't work, so it was something that we had to fake. We did pin a dummy lavaliere on the alien, so it appeared that the alien was using that mike, and the absence of mouth movement made the audience think the alien was speaking telepathically.

The actor's voice was recorded (without the mask) with slight reverb added. The metallic sound was created by placing the microphone in a metal trashcan (a clean one) and adding more high end (treble) with the equalizer. When the event happened, it was as if the great and powerful Wizard of Oz was back for a return engagement, complete with smoke, colored lights, and me behind the curtain.

#252 How to Muffle Outdoor Noises

Everyone enjoys the fall with the sound of leaves crunching under their feet. But if that's the only audible noise in the headsets, something must be done.

As a father and daughter walked along the edge of a lake discussing life's problems, the director wanted to hear their dialogue. Miking the talent from above, I used a zeppelined Sennheiser 416 shotgun with an extremely narrow pickup pattern (the same thing sleazy guys use in singles bars). As the camera tracked the talent, the boom would have been visible if they had been miked from below. This would have also eliminated the sound of the leaves. Because the Sennheiser was pointing down, it would be impossible not to hear their feet crunching the decaying leaves.

Tom Landis, the sound man, came up with a solution. He removed all the leaves from their path and had them walk on the grass. Since the camera was framed in a medium shot, the viewer wouldn't be able to see their feet.

Figure 32-3

When recording audio, don't leave the leaves alone—rake them up.

I recorded wild sound of just their feet traipsing on the leaves. This ambient track would be mixed with the actors' dialogue track.

This same principle applies to any extraneous sound that may be heard while recording dialogue. If possible, remove the offending noise, and record the dialogue. Remember to record an ambient track of that noise later and mix it (at a much lower level) with the actor's voices. Figure 32-3 should have told us that the leaves would have to be removed.

The same principle applies to bird noises when shooting in the woods. One or two birds in an ambient track are fine, but when 10,000 roust in a tree and scream, that's another story. Cannons and air rifles frighten the flock momentarily at shopping malls, but we didn't have access to a cannon. Any loud noise (an air horn, blanks in a gun, or running through the woods screaming and yelling while waving your arms) will do the trick temporarily. But after a few such episodes, the feathered creatures will get the hint.

#253 How to Record Audio in a Health Club

Sometimes recording audio at a location can be as difficult as recording video. Our health club shoot was no different in that we had to shoot and record audio during one of the club's busiest times of the day.

All the audio recorded in the health club could only be used as ambience because 200 treadmills were all operating simultaneously. If you're ever looking for health club ambience in a CD library, no other sound effect will do. The TV is always on, the radio is playing, people are exercising and talking, and some are even listening to their own music.

At our health club, each occupant was listening to music, chewing gum, sweating profusely, or running at top speed. When our actress turned and spoke to the camera, she was difficult to hear over the din of the machines. It was also very challenging to place a wireless microphone on her spandex Speedo. In her tight clothing, any freckle on her body would have been visible, let alone a wireless microphone and battery pack. The microphone itself was eventually gaffer-taped under one of the straps on her outfit and the wire attached to the receiver ran down her back. We gaffer-taped the receiver to her sneaker and told her to try not to walk with a limp; any other place would have been visible in the frame.

All her audio had to be heavily equalized because of the metallic thuds and pings of the exercise equipment. As she got up from her reclined position on one of the exercisers, she began to speak. Her voice suddenly dropped out because the microphone would slightly pull away from her body. Luckily, by riding the gain and equalization, I was able to find a happy medium.

Always try to record location sound rather than dubbing everything afterwards. The natural tone and ambience of the setting is important and would be lost if dubbing were your only choice.

#254 How to Remove Hiss

The best way to remove hiss is to kill the snake. If that isn't the problem, maybe you should continue searching for the source. I'll tell you a story when I was sound boy and we developed a case of the "hiss."

After closely monitoring the sound and making sure every take was acceptable during the shoot, the spot was soon in the can. But when editor Mike Gorga of Megcomm Productions heard the sound, I knew I would soon be in the can. A minute hiss had been picked up on the soundtrack that I hadn't heard on the headsets. This noise would be unacceptable in the completed national spot.

Hiss (analog or digital) normally runs across a pretty wide band of the frequency spectrum. If you are going to hiss, make it count. A sync blip was recorded at the head of the track to keep the sound in synchronization. Using the filters and EQ in the NLE system, the hiss still remained. The file was exported into Sound Edit 16 (16-bit sound) and the smoothing filter was applied to the bad sound. The smoothing filter took virtually all the hiss out but lost the voice presence, some of the actor's vocal range. This had to be replaced.

The file was reimported back into NLE and resynched. The vocal presence filter from the equalizer in the system was now applied. The better part of the sibilance had returned and made the voice sound too crisp again, but the hiss disappeared. The viewer could not tell that anything had been done to her voice. The other option would have been to loop and dub all of the audio, but that would have taken time from the extremely tight schedule.

This method does have some drawbacks, but they are minor. This smoothing and presence filter application should only be used as a fix, not something that should be done during the audio sweetening process. A percentage of the overall tonal quality was lost during this maneuver, but to most viewers this change will not be noticed; it certainly is better than hiss.

#255 How to Use Multiple Audio Tracks

When creating a video, often you will have numerous audio tracks expertly blended together into a final, happy mix. Instead of a mess of disjointed sounds, each track has a specific purpose and sound level that adds to the production.

When mixing the sound for a furniture commercial I did on the beach, I had to recreate most of the sounds in the editing room. It seems as if the 40-mile-an-hour winds, crashing waves, mocking seagulls, and crowds of onlookers made the audio unusable. A stereo mix of beach sound effects was obtained from the Aircraft CD (sound effects) because my sound picked up people talking and gulls complaining about our presence.

To this mix, random seagull sounds were added. I recorded the Foley tracks by watching the visuals and performing the tasks I observed onscreen. The two-channel Foley footage was digitized into the NLE system and added to the two channels of natural sound. These four channels were mixed within the system and down into two channels, each effect having its Foley counterpart and still remaining in stereo. These two mixed tracks were edited into a chronological sequence on the timeline: video track and audio channels one and two (stereo).

Through the magic of nonlinear digital editing, no quality was lost mixing the four tracks into two. The story was simple; through a series of close-ups we see that a businesswoman is having a really bad day. Things such as spilling her coffee on her report, copies getting jammed and shredded in the copy machine, her phone falling apart, getting a flat tire in the pouring rain, and traffic being gridlocked are examples of the visuals we see, as we hear her getting yelled at by her supervisor, the radio declaring her route home is closed because of an accident, and so on.

She gets home to find her key won't open the door, she hangs her coat up and the rack falls over, and she immediately crashes into her favorite sofa in her living room. The joke is then revealed as the camera pulls back to show the couch on the beach.

#256 How Tracks Should Be Broken Down

In the last solution, I discussed a beach shoot. This solution breaks down each track:

- **Tracks 1 and 2** These tracks were the stereo mix I mentioned previously of the beach sounds and Foley all in stereo.
- **Track 3** The voice over thought in the talent's head were slightly more complicated. In order to make the voices sound as if they actually were "in her head," reverb was needed to give it that echo quality. All the complaints (voices) were recorded to digital tape in a narration booth. This audio was also digitized into the NLE system. The reverb delay was created by staggering the sound by two frames. Channels 3 and 4 were used for this voiceover, with channel 3 being two frames ahead of channel 4. Played back, this simple echo effect created the voices in her head. Channels 3 and 4 were then mixed down to one channel (channel 3).
- **Tracks 4, 5, 6, and 7** The music was a stereo mix taken from the Aircraft CD music library. The stressful music was recorded on tracks 4 and 5, and the happy music (the music transitioned to) was recorded on tracks 6 and 7. As the mood of the spot changed from stressful to relaxed, the music dissolved into the calypso beat using the faders within the NLE. The audio levels on 4 and 5 would fade out to nothing as 6 and 7 would come into play. Luckily, a natural break in the music occurred when the stressful music changed to the happy music. Some can say this happens because of talent, but sometimes the music just works out perfectly the way it plays the first time. These four channels couldn't be mixed together because a stereo separation was needed for both pieces of music.
- **Track 8** The voiceover narration track was just a mono recording of a male voice asking questions like, "Are you having a bad day?"

After all eight of the audio tracks were in place, the creation of a layered mix was next. Obviously, the levels on the tracks had to be different; the more important ones had to be slightly louder than the others. In this case, tracks 1 and 2 (the natural sounds) were the most important. The music needed to be subtle enough to be heard, but not call attention to itself. The -20 level gave the music an almost subliminal quality; you felt it rather than heard it. The voices also had to be heard, but these were just one of a number of things that were going on in her jangled mind. Although slightly distorted with reverb, they were still audible; a -15 level worked best for the voices.

The inflated levels of the Foley sound effects made these tracks the most prominent in the mix. The enhanced sounds of the spilling liquid, the tearing paper, and the traffic noise made the viewer feel as unsettled as the actress. Once at the beach, these sounds disappeared and were replaced with calmer natural sounds like waves and seagulls.

With eight audio tracks in the system, another mix had to occur to reduce the number of tracks to two for the stereo videotape master. These tracks were mixed onto the left and right channels of the videotape, keeping the stereo elements (the music and sound effects) panned so they would remain in stereo for broadcast. Care has to be taken in the final mix to ensure integrity in the stereo. This mix was closely monitored with stereo headphones as well as the mono speaker on our color monitor. The mixing was done completely within the NLE with the left and right XLR outputs going into a Betacam SP deck. The sound effects, still the high-level sound, were simply louder than the other channels. Fortunately, no voiceover (in her head or narration) was overcome by the mix of the sound effects. It's almost as if these effects punctuated what she was hearing in her head.

Eight channels is a strange number for a mixed television spot. Some spots have over a hundred and some have only one. But in the case of this spot, the client loved it, the viewing audience got a chuckle from the humorous story line, and I proved to myself that you can make a creative, interesting, fun commercial with only eight channels. That leaves only 91 channels remaining in the NLE. The sequel to the spot must use at least a few more channels. Why would they put 99 channels there if we couldn't use them? Can you think of a better reason? I firmly believe that commercial audio should be seen as well as heard.

#257 How to Mike Gun Shots

Loud noises like gun shots are difficult to record because of the decibel level as well as the concussion afterward. Because it never sounds the same in real life as it does on video, sometimes the best bet is to use a sound effect from an established library. But if that isn't an option, I'll discuss how to pull it off without getting shot (like I did).

Working on the set of an HBO film called *The San Antone Kid* (this is the first time I've name-dropped), I was told to record "great sounds" for the gunfight segment of the film. The guns were old Colt .45s that were loaded with blanks so no one actually got shot in the fight sequence. Equipped with a shotgun microphone and headsets, I would record the firing of these pistols and edit them in after completion of the program.

A few things to remember about a shotgun microphone: It is the perfect choice for capturing sound because the pickup pattern is so narrow and focused. The other is that loud sounds in close proximity can damage the equipment. Don't put the mike right next to the gun or it will be destroyed. Instead, point it off to the side several feet away from the source.

Because of the heat and being in the desert, I wrapped my T-shirt around my head, clamping the headsets closer to my skull. Each gunfighter loaded their own gunpowder charges and used double loads because they make more noise. I got the sound perfectly from the cowboy in front of me, and the one behind me was going to fire next so I could record a different perspective of the sound.

Why he didn't point the gun in the air, toward the ground, or anywhere besides right at my back, I'll never know. When he fired, a flaming piece of powder shot from the barrel and hit me in the back, knocking me to the ground from the force. Still recording, I thought I had just digitally recorded my own death. I survived to tell the story, so keep in mind where your sounds are coming from . . . then step away.

#258 How to Record Dialogue

When people are speaking, try to capture the best sound you can in their environment. Looping or dubbing should only be used as a last resort, because it's not quite the same as location sound.

Place the microphones as close to the talent as possible. If using lavaliere microphones, make sure no clothing rustling is heard and nothing is blocking the pickup head. You have a little more leeway with a shotgun, and this pickup pattern has less bass frequency than a lavaliere. If the shotgun is pointed toward the mouth, you may pick up ventilation noises from the ducts overhead. The preferred method is pointing down to the mouth from above (if you have the room).

Another factor for good dialogue is to keep each person (if two are speaking) on their own audio channel. That way any equalization can be made to one without affecting the other. This also gives the editor more control in postproduction.

Make sure you record the entire line from beginning to end. Wait for the room to be quiet and listen for the beginning of the sentence. Don't stop until the director says "cut" to avoid chopping off any words.

If people are talking far away, their voices should be miked from farther away. If shot in a tight close-up, the sound should be closer and more intimate. Since video cameras don't make the noise film cameras once did, you will never have to contend with that sound problem.

In editing, let a few words of the dialogue precede the next scene or even let some of one person's words be heard over another's reaction shot. Every word does not have to be seen from moving lips. Keep the levels consistent and mike each source if possible. A mixer may be used if three or more people need to be miked and blended together. Label cables and microphones so you know in an instant which wire is connected to which unit.

Lastly, if you hear something that shouldn't be there or if you need another take, don't be afraid to make that point known. It's very difficult to come back for a retake at a later date and expect everything to match.

#259 How to Record Ambience

How can something that is considered nothing be so important? Ambience or room tone is the sound of the room without people talking. Most people don't know ambience is there until it's gone.

It is imperative to record ambience in every location, whether indoors or out. After the scene has wrapped and before people leave, record at least 30 seconds of this "nothingness." You might not need that much, but record more than you think you'll require. People told to be quiet have a tendency to snicker. Keep the audio level consistent with the dialogue. Don't raise or lower it to record ambience because that defeats the purpose.

No one, cast or crew, should leave the room before ambience is recorded. Each body produces or absorbs sound and that is important. Recording ambience in a full car or empty car at highway speeds produces different sounds. If you have a collection of ambience for every set or location, it will save the editor a lot of headaches.

Make it a practice for the soundperson or camera operator to call for quiet immediately after the location wrap. If it is part of someone's responsibilities, it won't be forgotten.

#260 How to Record Music

Often underrated, music is what sets the mood of a video. Whether it's recorded live, lifted legally from a CD, or assembled in the edit, music is needed to complete any project.

Some people question at what level music should be recorded. The answer: the highest level without distorting. You can always play the music back at a lower volume, but raising the level when you don't have it produces noise.

If recording a performance, mike as many of the vocalists or instruments as possible. Great books on recording music are available from `www.amazon.com`. These will show you how to correctly place a mike for each instrument and how many to use. *Mix Magazine*, a monthly publication, also runs features on recording music.

Listen to CDs of similar music and try to determine the correct levels for each instrument. Most professional recordings use a multitrack approach with sometimes 60 to 120 separate channels of sound.

When inserting the music into your video, a low volume is usually best. You don't want to overpower the dialogue and good music never calls attention to itself. It should help establish the mood. Raise and lower the levels at appropriate times and you will have the audience in the palm of your hand.

#261 How to Record Sound Effects

The simple way to get sound effects is to buy a CD containing a large selection. However, you may not find that specific effect you're looking for. A large postproduction house once called me looking for a ricochet sound effect. Either they thought I hung around with the wrong people or I had a library that was vast (a little of both).

Whatever specific sound you need, try to record it in several different environments. If it is a sound that cannot be reproduced easily (like a dinosaur), use your imagination and combine several effects to make a unique one.

In the original *Star Wars*, most of the sounds needed didn't exist in anyone's sound libraries. When they wanted a laser blaster sound, they took a crescent wrench and hit a taut, metal guide wire. The "twang" noise was processed and slowed down, providing a new effect (which you can now purchase in sound libraries).

In the field, try different microphone placements to capture the sound. Try from below, above, and to the side. Mike the sound close as well as farther away. In postproduction, what happens if you speed the sound up slightly or slow it down? Sometimes playing a sound backwards will add a unique perspective. Play with echo, flange, equalization, and reverb, varying the decay rate until you achieve what you're after.

Chapter 33

Choosing the Best Microphone

#262 How to Determine Which Microphone Works Best

In this chapter, we'll discuss the pros and cons of shotgun and lavaliere mikes, using boomy sound as an advantage, taping into a sound system, using a wireless microphone, getting the best level, silencing appliances, dealing with battery failure, recording pets, recording sound outdoors, where to place microphones for the best pickup, and how to hold a boom comfortably.

#263 How to Use a Shotgun Microphone

There are times when a shotgun is the perfect choice of microphone and may be the only one that will work in a given situation.

On a documentary at a Christian rock concert, Tom Landis, the shotgun wielding soundman, and I had to travel, via golf cart, over 600 acres and through 45,000 campers to find out why they attended this three-day event. The shotgun was a Sennheiser 416 loaded in its zeppelin and ready for action.

As we wandered through the campsite, we would walk up to people, interrupt what they were doing, and ask questions. Either Tom or I would introduce ourselves and explain our mission. If they wanted to cooperate (they all did) we'd have them sign a release form and begin shooting.

Not knowing what to expect in sound level, or if talking with a group—who would speak first, Tom really had to be on the ball. If one person was speaking and another would have a thought, the shotgun would be pointed in their direction. The pick-up pattern is limited and unless pointed at the mouth, the level and clarity is lost. In fact if someone walked up to Tom from behind, he couldn't hear them approach unless the shotgun was pointed in that direction.

If we had done the same thing during the concert, the shotgun could have picked out specific voices in the crowd even when a band was performing. This selectivity is what makes the shotgun excel in these situations. Like a parabolic mike used for surveillance, you can now selectively hear only what you want to hear.

Keeping one eye where the mike is pointed, the other at the VU meter levels on his mixer, and if he had a third, it would be watching where I was so the mike wouldn't get into the shot. But a professional knows his limits (how close the mike can get to the frame) and Tom instinctively kept a safe distance.

Much like a rifle with a site, when you point a shotgun mike at something, you know it will pick it up, the level at which you do that is your main concern. There is no magic way to accomplish a shoot like this: just be prepared for anything, point the shotgun at what needs to be recorded, and ride the levels to get the best mix. It's as easy as baking a 12-layer cake—with no icing.

#264 Review: The Pros of Shotguns

- With a shotgun microphone, people never have to wonder where the microphone is. It's big and they can see it, so they'll want to hold it and talk in it (much like a Popsicle on a hot day).
- This type of microphone will only pick up sounds in front; any noise behind will not be heard. So if your wife is yelling at you, point the mike away from her and you won't hear anything.
- If several people are sitting in a group and you want to pick out a particular voice, this is the mike of choice. Think of it as a toy gun (a real gun gets you in trouble). Wherever you point it, that's what you'll hear (unless you drink too much and you'll be hearing other things too).
- This type of mike isn't attached to clothing so they'll be no rustling (cattle people beware) or muffled sound. Clean sound will be recorded (if that's what you want).
- If you only have one microphone and several people in a group will be talking, a shotgun microphone may be pointed at whoever's speaking. This gives the boom person something to do.

#265 Review: The Cons of Shotguns

- If you are a snitch and want to be wired for sound, this isn't the mike of choice.

- If the talent is moving, you must stay with them at a constant distance to record the sound. When they start running and ducking into back alleys, you may wish it was a real gun.
- Because the pickup pattern is so narrow, a turn of the talent's head might send him or her into the dead zone (not the movie, just where the microphone can't hear it).
- Since the microphone is out in the open, it must have protection (from the wind or Mafia). Wind noise, although somewhat controllable, is more of an issue with this tool.
- You may not always be able to get close enough to the talent for those romantic scenes. Sticking a shotgun microphone where it's not wanted will cause problems.

#266 How to Use a Lavaliere

Another trick you can do with a lavaliere microphone helps hide it from sight, but also allows you to record sound without clothing rustle—only if you are a woman. It is sexist, but the principle you are about to learn will not work for a male—not without surgery.

On a recent shoot, the female actor had a lot of speaking on camera, and her turtleneck sweater offered no place to hide the mike without it brushing against clothing. Our sound man, Dave Wilson, had a solution that saved the day. The lavaliere would be taped (with surgical tape because it doesn't stick around too long) between the actor's breasts—in her cleavage.

Although it sounds rude and intimate, have the woman run the mike under her shirt and attach the mike head to her bra, between the breasts with surgical tape. Her cleavage will prevent the clothing from touching the mike head and the surgical tape will prevent it from wandering around. Gaffer tape is too heavy and masking tape is not strong enough. When you think about it, it really is a great idea. The wires are nowhere to be seen. The mike's head is close to where the sound is emanating and invisible, and the tape is attached to the bra instead of the skin.

Have the actor perform the attachment in a private area. If you have the tape already attached to the mike head that will save time. This will work with any woman—just make sure she is wearing a bra or you will have other problems to contend with.

#267 Review: The Pros of Lavalieres

- This type of microphone can be hidden, so no one will know you are wearing a microphone (unless you tell them). Like the emperor's new clothes, this is an invisible mike.

- Since this microphone is attached to the person, you will be miking them closer. Sometimes this is the only type that will fit (I won't go into the mikes you swallow).
- A person can move around while wearing this type of microphone and the sound level will stay constant. Talent does sound different when they reach the end of their cord.
- Because of the wider pickup pattern, two people sitting side by side can sometimes use this one mike. It will pick up sound from most places (just make sure you wash your hands; you don't know where it's been).
- On interviews, the lavaliere will pick up your questions as well as their answers. Throwing a microphone back and forth between interviewer and interviewee doesn't work.

#268 Review: The Cons of Lavalieres

- Because these mikes are hidden and worn close to the body, clothing rustling is almost always present. When doing nudist videos, clothing rustle is not a problem, but hiding the mike is.
- Since the mike is attached to a person, his or her movement might be limited. You should try having a metal or plastic orb connected to your body and then be told to "just be natural."
- If two lavalieres are placed close to each other, phase cancellation may occur. This can be adjusted, but once they begin fighting, they are never the same again.
- These microphones are more fragile than others because they are sat on, twisted, yanked on and off, blown into, pounded on, and, in an exercise video, perspired on. Be glad you have your current job; the life of a lavaliere is no fun.

#269 Review: The 10 Most Important Ways to Record Sound

1. Use the right type of microphone in the right environment. Don't use a shotgun in a situation that doesn't warrant its use and don't use it as a lightsaber.

2. Whenever possible, hide the microphone so it doesn't call attention to itself in the spot. Although they don't speak, they are better to not be seen and always heard.
3. Make sure the person wearing the mike is comfortable with how it's attached. Don't use nails and don't tape it to his or her skin.
4. Use a windscreen when recording sound outdoors. Even if you don't have gale force winds, the acoustically transparent covering protects the mike's elements from nature's elements.
5. Record all sound at a decent level; you can always lower the volume in editing. Muddy sound only gets your shoes dirty.
6. Always record without filters or equalization in the field. These elements can be added in postproduction if needed; they can't be removed easily (like that wart before your big date).
7. When recording sound effects, record as much as you can as loud as you are able. It may be hard or impossible to recreate certain effects (that rhino's tonsillectomy is a story I should tell you about sometime).
8. Try to keep each different effect on its own audio channel. This allows more freedom when mixing in post (it also stops fighting).
9. Keep everything in stereo in each level of your editing process. Sometimes having sounds coming out of one side of the room will add to the effect's dimension (people may call you one-sided, but that's their problem).
10. If the live, recorded sound doesn't work for whatever reason, don't be afraid to recreate, redub, or remove it.

Look at other people's work (don't copy it; look at it). How did they achieve a certain effect? What worked and what didn't? I don't think that you'll ever run into the situation where you are using too much sound on a video. Every effect has its place when used correctly.

#270 When Boomy Sound Is Acceptable

Whenever I'm asked to record sound on a shoot, I want it to sound exceptionally good because people will yell at me if it doesn't. On a recent shoot, I recorded sound and did my best to make it sound acceptable.

The actress was to descend a staircase and walk as she talked (even I can do that occasionally). The first microphone option was to use a shotgun on a fishpole and follow the talent from above (the balcony, not heaven). We chose to use a Sennheiser 416 shotgun mike for its accurate portrayal of sound. The mike's pickup pattern is as narrow as a flea's eyebrow, so the unit must follow the source of the sound.

During rehearsals, the talent's voice sounded very hollow and boomy because the foyer was two and a half stories tall. I was miking from above because I would have blocked the camera if I accomplished the task from below. In my lofty environment, this endless volume of drywall space made any noise full of echoes and hollow sounding. Unfortunately, this was the nature of the beast. I could have covered all the vertical surfaces with sound blankets or sound soak, but that would have been visible to the camera.

Some of this hollowness was expected because the camera was shooting up at the talent as she descended the staircase, and the viewer would notice the cavernous hallway. Since the set was tight, enclosed, and confined, that type of sound would not be objectionable. The eye rarely lies; we expect to hear a certain type of sound when we see the room. Therefore, a little hollowness should be expected.

#271 How to Find a Sound System

How do you tie your microphones into a group presentation if you can't access the sound system? This happens more than you think. When you're asked to record a presentation of several people (I had eight people), try to tap into the house system. If you can, you won't have to add an additional microphone to every person just for your video source. Most people don't like to have two mikes when they're presenting. It gets cumbersome when they have to switch it on and off, and no one likes to address the phase issue (if 2 mikes are exactly the same, they may cancel each other out).

I usually just get output from the house's sound system and patch that directly into my camera. I always travel to these places with a bag full of attenuators, phase reversers, adapters, cables, and connectors. You never know what type of system you'll run into, and it's good to have your own adapters to make everything fit.

The last presentation I attended was unusual. The house system had no audio output. Actually, it had one, but the house sound engineer disconnected the audio output inside the unit. He didn't want anyone recording sound from his system, so he made it impossible to use the normal output. I don't know if this was job security, jealously, or simply the first sign of madness.

When you encounter a situation like this, you must have a backup plan. You don't want to rewire the system and you also don't want to attach eight different microphones to all the talent. The client also wanted the audience to ask questions via handheld wireless mikes that would be passed to the person asking the question.

We now had 12 sources of audio to deal with. The only outlet (without using a boom mike and getting distracting background noises) was to use the house's sound system. I gaffer-taped a lavaliere microphone a few inches from one of the speakers in the ceiling. This way I could still use the 12 mikes in the house's system and also get a clean feed. I knew that this looked cheap and amateurish, but sometimes the sleazy method is the only way to capture the best sound.

In the end, the client was pleased with the sound and we didn't incur the wrath of the system's engineer. Always carry spare equipment and connectors so you can be ready for most circumstances. Without a doubt, you will encounter some very unusual setups. When you do, look for the most logical recourse, and when that doesn't work, cheat.

#272 How to Use a Wireless Microphone

To go wired or wireless depends on your given situation, equipment, and budget. Wired mikes are more cumbersome, and everyone has heard the story about the talent forgetting to turn his wireless mike off when using the bathroom. But if you really have the choice, the new wireless microphones are tough to beat.

Wireless microphones can be used anywhere because you can let the actors move as they want without worrying about tripping over wires. The only problem I've ever experienced with a wireless was picking up some disembodied voice that wasn't the actor's. But with UHF multichannel capabilities, this rarely occurs anymore.

On a recent shoot, I had a chance to try out a new wireless package. The transmitter unit would slip innocently in the talent's back pocket, only allowing the tiny antenna to dangle out. Powered by a 9V battery, we shot all day without the battery wavering. The transmitter was connected to a miniature cardiod lavaliere microphone that was one of the smallest I've seen.

The receiver is attached by Velcro to the back of the camera and is powered all day by four AA batteries. With two flexi antennas (A and B) and some BNC mounts on the receiver, we essentially have two channels, but the helical filter really suppresses most unwanted signals. I got as far as 200 feet from the base unit and I could still hear every word clearly. When doing research before buying a mike, I learned that most operate in the range of channel 57 to channel 59. The *radio frequency* (RF) range is safer to use there because some wireless units actually travel higher in the UHF channel range (67 to 69) where DTV stations may someday reside.

#273 How to Use a Wireless Microphone with Numerous Actors

Did you ever try to get party people (all over 60) to wear paper hats? Our shoot involved a birthday party with seven guests. The birthday lady, wheelchair-bound, would roll into the kitchen as her friends yelled "happy birthday" (what do you think they would yell at her age?). The TV spot, one of the few 60-second productions I've done, would highlight the versatility of her mobile chair and how her lifestyle wouldn't change now that she got jazzy.

Because the shot was extremely wide and we needed to see the entire kitchen, a boom microphone was out of the question. Seeing the entire room allowed no hiding places for a directional, shotgun microphone. Instead, one of our wireless units was taped under the table to capture the cheers as the woman entered. Although a cardiod, the close gathering of these people didn't necessitate an omni-directional.

She was miked with a second wireless lavaliere. A wired microphone's cables would have been visible in the shot and may have tangled with the motorized wheels. With the windows gelled with Rosco 85ND9 and the Arri 1K and 650-watt tungsten lights illuminating the room, the only visible flaw in the scene was the three suspended halogen lights over the kitchen table. After Tom Landis carefully cut a piece of Tough Frost gel and gaffer-taped it over the glass surface, the halogens were less piercing on the white birthday cake.

As the woman enjoys her party, she reminisces through voiceover narration. Since we didn't want her lips moving as she was recovering from the shock of her surprise party, we recorded her voiceover at a different time using the same lavaliere microphone.

In another area of the beautiful, cherry-stained home (kids did eat cherries and left stains everywhere), we staged a medium shot of the woman in her wheelchair and recorded her "testimonial." We also recorded her visually because different portions of her scripted testimonial would be appearing in other spots. We could either lift just the audio or both depending on our need.

The hardwood surfaces of the home proved to be a problem even with the close proximity of the lavaliere microphone. Her sweet voice echoed off all the hard, vertical surfaces. Our only solution was to cover every surface in the room with sound blankets in the hopes that it would absorb some of the harshness.

Twenty thick, scratchy wool blankets later and we had a set that looked like Hannibal's cell in *Silence of the Lambs*. If the actor would have said

"Clarice" just once, I was history. We gaffer-taped the blankets to the walls, suspended them from C-stand arms, shoved them under her wheelchair, and covered the camera; I was allowed only a miniscule breathing hole. If anything in the room looked hard and sound reflecting, it was covered with a blanket (the owner's pet ferret wasn't too happy about that).

With the room deadened, the woman spoke her lines and we recorded it with the clarity we hoped we would have. Once the wireless mike was placed on her body and hidden, she wore it the rest of the day and didn't have to worry about it. Wireless microphones really take the headache out of audio recording. They have no cables to wrangle and no XLR connectors to get bent, driven over, or broken by frustrated grips. As long as you have power, you can go anywhere (I know a few dethroned kings who used to believe that).

#274 Review: Five Reasons Why Wireless Is the Only Way to Fly

1. Wireless microphones are really more portable than wired units. The actors can move around to their heart's content without tripping or getting tangled in the wires. Of course, you can no longer play the "let's tie up the talent" game anymore without the mike cables.
2. With several channels, filters to screen out extraneous crap, and a receiver close to the audio person, you are still in constant control of your environment. How often can you say that about anything else?
3. The close proximity of the mike to the talent means sharper, clearer sound and it allows you to record sound where shotgun mikes would be visible to call attention to themselves. If you're ever asked to intern on a shoot and need to hold the mike for long periods of time, volunteer to hold the lavaliere.
4. Constantly monitor the sound of the wireless unit. The talent may forget they are wearing it after extended periods of time and it's up to the discretion of the audio person when to turn down the volume on the receiver. Of course, local CB operators will still get a cheap thrill, but that's all they usually ask for anyway.
5. A wireless microphone should be treated as any other mike. Extraneous noise, hard- or dead-sounding rooms, and microphone placement are still just as important. Change the batteries daily (like your underwear), protect the mike's element (just like your underwear does), and don't have it pinned to the talent too long (just like your nametag on your underwear).

#275 How to Get the Best Sound Level

Part of the magic of being a freelancer is that you never know what tomorrow will bring. Little Orphan Annie sang about it, but we video people don't know what to expect.

You can get a call anytime that requires you to be on location for a production. Your gear must be mobile and you will need everything you own. Like the "one-man band" of the past with cymbals on your knees, harmonica in your mouth, and bass drum on your chest, you become a video professional.

I got a call from *Miracle Pet*s of Vancouver about a shoot that was happening in my area. *Miracle Pets* is a one-hour program on the PAX Network that highlights amazing things that pets have done. They were sending a producer and a shooter from British Columbia and needed a sound person (me) to help out.

The story we were about to recreate was nothing short of amazing. A woman, Tracy, and her show dog, Bo, arrived on the scene of a car accident and immediately helped by directing traffic. I don't believe the dog did anything but blow a whistle and wave his paw; the woman did most of the work. Suddenly, her dog left the incident and raced into a nearby field. Running after Bo, she discovered a man who had been thrown from one of the cars lying in the field, bleeding heavily. Bo, hearing the man's cries for help when no one else could, actually saved him from bleeding to death (Bo didn't know CPR, but he called others who did). No one knew this man had been thrown from the car and the driver was still unconscious.

I met the Canadian crew for the first time at Tracy's home for the shooting of her interview. The crew had been traveling for three weeks on numerous shoots and was hiring local audio help in the specific areas.

One of the best things about this business is the immediate bond you have with the other crewmembers. I had never met or spoken to the crew before, but it was like I had known them for years. Tracy even commented at the end of the day that she had thought we were a well-oiled machine (my hair was greasy) and assumed we had worked together in the past. The bond that forms among crewmembers is unique; they quickly become a family and often get along much better than most families do.

Each person on a crew is given a particular role and they are expected to do the best job they can. We are all professionals and if hired as a sound boy, I will do that job to the best of my ability. Now my saga may begin.

#276 How to Silence Appliances when Recording Sound

The camera operator set up lighting for the interview in a small, cozy room in Tracy's home. As the soundman, I was hidden in a miniscule alcove leading to the basement. Unlike a small child, I needed to be heard and not seen. Tracy's voice was recorded with a shotgun clamped to a C-stand on her left and an omni-directional lavaliere carefully clipped on her shirt. Both microphones were attached via XLR cables to a three-channel mixer. A stereo feed was then sent to the camera.

Another thing about being prepared (you can never find a Boy Scout when you need one) is to bring everything you may need to the shoot. I brought a shotgun, wired and wireless lavalieres, a mixer, boom, cables, connectors, and batteries. Not knowing exactly what the producer would want, I had to be ready.

A soundperson's equipment needs the ability to fit any camera, supply tone, or a mixed signal, and they require numerous other things to do their job correctly. Every interviewer's style is different and you must expect the unexpected, like battery failure.

#277 How to Deal with Battery Failure

My microphones were powered by batteries, as was my mixer. Running on DC for extended periods of time will suck the life out of anything (just ask my mother-in-law). In the middle of an extended take, my battery indicator on the mixer started flashing. This is something that must be called to the attention of someone before the sound ceases to happen. In my foxhole, I had a spare set of batteries (three nine-volt) and had to change them within a millisecond, knowing everyone was waiting for me.

I never like stopping a take. If something falls over or explodes, others usually see it and will stop the action. When the soundperson says, "Hold it!" he rarely wants someone to hold the microphone. Because the problems I encounter are heard and not seen, I must explain why I have to stop.

#278 How to Use a Microphone with Pets

The life of a soundperson is never easy. If a problem ever occurs with extraneous noise on a shoot, everyone looks at you. When you are all set up and don the headphones to listen carefully to your environment, you must pinpoint every sound and isolate its source. I heard eight fans, three appliances, two barking dogs, and a canary with the hiccups. It was then my job to jump up and silence everything that was making noise.

I've turned off a million refrigerators in my day (by talking sleazy to them) and remembered to turn half of them back on (if you can't turn on a refrigerator, you're in trouble). The cameraman told me something that he claims is an old soundperson's trick (I didn't know about it maybe because I'm not *old*). When you turn off the fridge, put your car keys inside. You will not go very far without them and it will remind you to restore power before you leave.

When one of Tracy's other dogs, a three month old named Ty, began barking in the basement, everyone once again looked at me. It wasn't my dog, I didn't make him bark, but I had to tell him to cease. A garbage truck would lumber by, a chipmunk in the garden would have a conniption, and the telephone would ring all at the wrong time. When the stares turned to me, I realized I had become known as "the Silencer." The split second I heard the offending sound in the headset, everyone else heard it too. Being in charge of sound, I had to locate someone in the Mafia and have that offending something "whacked."

#279 How to Record Sound Outdoors

I knew once we shot outside I would have more control over my environment (remember I'm young and naive). I love to shoot because as soon as you point the camera at anyone, they immediately freeze and give you the "deer in the headlights" stare so you can do your job. When you point a shotgun mike at a dog, they walk up and sniff it. The only thing worse than hearing sniffing that close is when a dog decides to devour a bone by sucking the marrow from its center.

As Tracy and Bo reenacted their scene outdoors, I miked her with a wireless lavaliere. How do you disguise a lavaliere on someone wearing a low-cut tank top without going where no man has gone before? I used a thin strip of gaffer tape under her strap and snaked the wire underneath. Having no pockets in her shorts for the transmitter, where do you hide this battery-operated device where it will not bulge?

When you must hide it someplace you're not allowed to go, have them perform this task. Remember to tell your talent to clip it rather than just slide it under the clothing. When we shot the jogging scene, she knew why clipping was so important.

Being ready for almost anything, I had enough XLR cable to record sound where the *Titanic* lay at the bottom of the Atlantic. Using the shotgun for ambient sound in addition to the wireless, I had cables going from the zeppelin and mike to its power pack, from the pack to the mixer, and from the mixer to the camera. The wireless receiver I stuck in the mixer's Portabrace. My headsets were connected via a three-foot cable, meaning I had to wear the mixer around my neck to avoid looking like I belong to one of those tribes that stretch their neck with plates. If I wandered near water, I knew I would never surface again.

With everything neatly coiled and tied in front of me, I gently placed my fishpole and boom on top of my spare cables as I sent tone to the camera. At that moment, Ty wandered over to me and began licking my hand. I thought it was nice of him to show affection after I had told him to shut up a few hours earlier in the basement. It appeared he held no grudges because he was so young and still a puppy.

With a sly smile, he walked over to my shotgun mike and lifted his leg. They say that people two miles away heard my scream. As I threw myself over my expensive microphone, it seemed Ty had gotten his revenge (try explaining those stains to your wife!).

The rest of the day went without incident and I made two new friends in this illustrious business (five if you count the quadrupeds). When you are hired to perform a particular task, you're expected to know what to do without having to be told. This comradery between video people exists throughout the world. I've worked with crews that didn't speak English and yet I understood what I needed to do (besides learn their language).

So if you ever need me for one of your shoots or desire talented individuals who will perform their tasks without complaining, just give me a call. On second thought, maybe you better *not* call me first. When asked to be a soundperson, I now always wrap my equipment in plastic because of the "dog in the grass" incident.

#280 Review: Eight Things Some Will Try to Forget

1. Be willing to wear many different hats in this business. If you are asked to grip on a shoot, that role is just as important as the director's. Just being part of a crew is exciting to me, but I'm also the one that makes everybody wear hats at a birthday party.
2. Bring everything you may need to the shoot. You might not know the specific situation until you arrive, so being prepared will avoid embarrassment. When I walk in with a case containing 12 microphones, it at least keeps the rest of the crew quiet for a few moments (some are woodened and painted to look real).
3. Bring enough spare batteries to power a small starship. Batteries will die when you least expect them to, so have others with you. I did get dirty looks when I backed my car into the kitchen.

4. Have enough cables but also have short lengths. Longer cables may be coiled, but when you're on the run, they tend to loosen (just like my shoelaces when I was chasing Ty after he tried to water my microphone).
5. Be ready to attach your sound equipment to any type of camera, and bringing a mixer is mandatory for tone generation (no relation to *Star Trek: The Next Generation*). Once the signal into your mixer is set, make sure the one going into the camera is also happy. Just because it sounds okay to you doesn't mean the camera's input isn't over- or undermodulated. This is one of the few times you will be totally in charge. Don't foul it up!
6. Good headsets will not allow you to hear anything else but what the microphones record. If you hear obnoxious noises in the headsets, make sure they aren't just filtering in from your surroundings, like when the puppy began to . . .
7. Remember to turn on anything you silenced when recording audio. When you tell a dog to be quiet, do it nicely because they will enact revenge when you least expect it.
8. Make yourself as mobile as possible. When the camera operator begins to move, you will follow because your leash is short. No matter where I work, everything seems to have a reference to dogs.

#281 How to Place Microphones for Optimal Pickup

The placement of the microphone and the recording level are the two most important aspects of recording sound (besides the operator's skill, the type of microphone, and so on). The mike needs to be as close to the source as possible without interfering with the camera or the action. If the sound is recorded too high or too low, it will be difficult to make it sound good. Figure 33-1 shows how the locomotive dwarfed the crew.

When having a team of two people on sound (a boom operator and a DAT operator), both people must have headphones. If you only have one headset, it should be used by the operator, but you'll be much happier with the end result if you have two headsets.

It doesn't matter if you are recording the sound of a cricket sleeping or the sound of a Titan missile; both parties need to be able to monitor the sound. If only the operator is wearing a headset, he or she has to signal the boom operator to move closer or get farther away. All this waving attracts attention to the sound people. Remember, sound should be heard and not seen. If the boom operator is wearing a headset, that person can judge if he or she should be closer or farther away. Turning up or down the volume isn't

Figure 33-1

1909 technology versus twenty-first century equipment

the same as moving closer to or farther away from the source. A distance difference exists and the viewer will hear that. The toughest set of ears you will ever find is that of the inexperienced listener.

Now some people say they can't afford two headsets on a limited budget. Yes, it does cost more to have two headsets, but the people who are hiring you will get a much better product. If your deck only has one output, buy a splitter or Y cable so it can accommodate two headsets.

You also shouldn't scrimp when it comes to headsets. The professional models do make your head pointed and muss your hair, but they keep out external noises and let you hear a clearer signal. My rule of thumb is, if the headsets are on and the mike is live, you shouldn't be able to hear someone talking next to you; the headsets will block out that noise. If you invest in a $2 pair of ear buds, you will have no idea if the mike is picking up the sound or if it is leaking through your ears. All you'll get is some nasty stuff coating the ear bud foam.

#282 How to Hold the Boom

Another difficulty is finding a comfortable position to hold the boom for long periods of time. You want to hold the boom as far back as comfortable. If you choke up on the boom, any movement you make will telegraph onto the mike. Because I'm slow, I sometimes have a hard time telling if the mike if facing up or down when it's in the zeppelin. An easy way to remember is to attach the mike cable to the top of the fishpole with a hair braid. Hair braids are those little pieces of elastic with brightly colored plastic balls on the end. If the braid is facing up, so is the mike. This braid keeps the cable neat and if someone on the set wants to wear a ponytail, you can make another friend.

Should you hold a boom over your head or against your chest? Which would you prefer? It doesn't matter as far as the sound is concerned, but the higher you hold the boom away from your chest, the stronger your arms must be. The key thing to remember is to find a comfortable position because you may be in that position for quite a while.

Mike placement is also important. You need to get as close to the object as possible without the camera seeing the mike in the shot (or in our case, getting crushed by a moving train). Lower the boom into the frame and have the camera operator tell you when he or she sees it. Then give yourself at least a foot of leeway. Some boom operators have a small video monitor attached to their belts so they can watch a video feed. Our budget didn't allow for that, but it's a nice luxury if you can afford it.

The presence of a shotgun is unequaled. Most extraneous noise is not recorded because of its limited and narrow pickup pattern. I prefer the closeness of sound with a boom mike. That closeness is also found with a lavaliere, but you also have to hide that and usually get clothing rustle. A shotgun enables you to get great sound at an invisible distance.

When working with a shotgun and boom, remember you are tethered to your sound partner with a cable. Give yourself enough leash and neatly coil the rest. The director tripped over my cable and glared at me for the rest of the shoot. One solution is to hang stuff from the belt loops on your pants. You'll be carrying cables, so make the belt loops serve their purpose. The key here is neatness. Neat is faster and you will be expected to move fast.

If miking from above doesn't sound acceptable, try miking from below. You can still keep the mike out of the shot if you reverse the placement. If your cabling happens to fall on the ground, make sure it's 90 degrees rather than parallel to electrical cables. Cables that aren't crossed might produce an AC hum.

With headsets, you will hear all kinds of extraneous noises. Use your boom as a makeshift metal detector and search out where the offending noise is located. A ticking sound isn't always coming from a watch or time bomb.

Try to get the best possible sound out in the field. Some people are lazy and believe that anything can be fixed in postproduction with Pro Tools. Pro Tools is amazing, but it won't fix sloppy audio. It's easy to recreate the sound of a train, but it's far better to record clean sound of a 1909 steam engine as it moves by the microphone.

Hopefully, a lot of things will be common sense. Always carry extra batteries for the microphone and field recorder. Have gaffer tape, black wrap, needle-nose pliers, jeweler's screwdrivers, and electrical tape with you at all times. Although you may not be a grip, you might need to make repairs or quick changes on the fly.

Other larger items like blankets are handy to muffle unwanted noises. Just remember to replug any noisy object that you unplugged.

The best thing you can do when recording sound is to expect the unexpected. Most times on a shoot, problems are usually of a sound nature. Determine the problem and try to find the best solution.

The best advice I can give you is to practice with the headsets and boom. Every environment is unique and with a little training you can make Hollywood look better.

After I recorded the sound of the steam locomotive, I will no longer volunteer to stand on the tracks alone.

Solutions in this Chapter:

Chapter 34

Postproduction Challenges

#284 How to Edit

Editing is where many projects are made better because of talent in the editing room. Sometimes mistakes can be corrected, shots made to look better, and storylines strengthened.

In this chapter, we'll talk about how to make a statement with editing, how to do your editing on paper first, how to layer effectively, how to edit music and sound effects, and different options available for deliverables.

#285 How to Make a Statement with Editing

Editors have a lot more say than they think. The best shot or scripted video can look like terrible if not edited correctly. If you want to make a statement by cutting a spot a certain way, do it.

You will be establishing the pace for the video and your skill as an editor can transform the project. In preparation, watch other videos for how they were edited.

Remember to establish a rhythmic pacing and stick to it for the duration of the spot. My boss used to slap me on the back at every cut point. Edit as tightly as possible. Keep each shot on the screen only as long as necessary to get the point across. Change the speed (running time) of a shot slightly to extend it or shorten it. See Figure 34-1.

Figure 34-1
An editor in his environment

#286 How to Do a Paper Edit

In order to make a large editing project more palatable, break the complicated edit down on paper. A paper edit can be redone a thousand times without editing any video footage, renting equipment, or incurring too much expense because it all happens on paper.

The following process works best in documentaries with an interview. The first step is to have every single word of the interviews transcribed on a computer. This is the only way to make sure the best pieces are used without wearing a hole in your tapes. If someone has twenty thousand "ahs" or "umms" in their first sentence, that could be noted on the transcription. This is where word processors really shine. Paragraphs, sentences, and words can be rearranged, manipulated, and edited to suit your needs.

With all your dialogue transcribed, use the nifty highlighting feature of your word processor and underscore the sound bites that make the most sense. I used to highlight each line with a yellow marker, but my mother thought I was jaundice and lacking something in my diet.

Have the highlighted scripts reviewed by everyone who needs to see them. This way the client can't come back later and ask why something wasn't said by someone. Everything is in front of them in black and white (and some highlight color). This step also gives you a rough idea of the size of the rough edit.

When all the best footage and takes have been selected, highlighted, and put in the correct sequence on the paper edit (cut and pasted), you can safely move to the capturing stage for editing. Fifty reedits at the paper stage are more tolerable than the same amount when in an offline or online session.

#287 How to Use Layers Effectively

In order to stand out from the pack, you need your videos to be unique. One way to get the viewer's attention is with multiple layers in your video. I don't mean shooting on thick videotape, but using more than one layer in editing. You have an almost unlimited supply of video layers in most *nonlinear editing* (NLE) systems, so use this free tool and add depth to your next production.

Now that we all agree that you want your work to look better than the rest, just how do you accomplish that? One easy answer is layering, the same thing you're supposed to do with clothes when you go outside in the winter. But if you just put 20 things up on the screen at the same time, it's just going to look confusing, not better. This is where the second special method comes in: movement. So now you have moving layers (sounds like a '60s rock group).

This isn't a new concept. You've all seen it done before. It enables much more to be seen in a given frame and the human brain likes that (some clients do too). This technique has been done in audio for a long time; people are just finally getting around to it in video.

The best place to start with this novel creation is on paper. Never go into the editing room and just do it (not unless you work for Nike). The client must approve what you're doing, and odds are they have no idea what a layered, moving video is.

I had a client that wanted a weekly commercial for his food store. The way he always wanted the spot done was nothing special. A catchy jingle opened the spot, three products were offered at a sale price, and the location of the store would be displayed. A friend of mine at a post house wondered how long I could keep on producing these "cookie cutter" spots. I was tired of doing the same thing all the time and the client actually wanted a change (that's a good thing).

We then created the new multilayered moving approach. The client wanted an opening that could be used week after week, but the products-on-sale segment would have to be much different. After convincing him not to have the cheesy flying box (a box with a video image that moves quickly across the screen) special effect open the commercial, we compromised on a well-lit, colorful opening.

The opening would really catch the viewer's eye. It was something you wouldn't expect to see in a grocery store commercial. A tabletop display of the freshest fruits and vegetables was created. We didn't use the usual grapes, apples, or lemons, but instead used much more exotic, colorful fruit like star fruit, cut sangria (a type of watermelon), watermelon, cherries, and so on. Most of the fruit we used I couldn't pronounce, but it sure looked colorful.

The display was kept attractive by constantly adding water to the edible stars. A garden hose with a fine mist spray was hung directly over the display. The tiny droplets of water made the fruit and vegetables look appetizing to the eye. Obviously, we had to do this outside because our studio didn't have flood insurance, but with streaks of color-corrected light on the products, the highlights would make the viewer hungry for more. To create a "moving" motif, a helper would tumble blueberries from above, off screen. This blue movement made it look as if it were raining blueberries. We wanted a deep depth of field, so shooting outside at F11 really helped.

This opening then dissolved into the "donut" or weekly changing product segment. We started by shooting the products on a neutral background. Potato chips and other bags of goodies had to be propped up, but it wasn't difficult to hide the wooden blocks behind the bags. For the finished effect we wanted, it was imperative that the camera was locked down. The subject was modeled with lighting, but the camera couldn't move.

The digital tape was then taken into editing. There the products were "grabbed" and manipulated. The entire background was removed from behind and around the product. A new keyable background (zero black) was created. This super black background would enable us to key the product over anything.

We then generated a moving background with the client's logo moving from the lower right diagonally to the upper left over a textured, 3-D-like background in the NLE system. This "endless-loop" background would be the base layer.

The prices were created as character-generated text. We now had all of our layers in place. The next step was to put them all together.

Layer one was the moving background (remember the word moving). Layer two was the product on zero black. This image expanded and moved vertically over a 15-second period of time. Because the product would only remain on the screen for five seconds, this longer movement period allowed it to grow (or move) very slowly. The product was layered over the background.

Layer three was the price text. These CGs were placed on their own layer and made to move horizontally and expand between the letters. This movement was programmed for 20 seconds because we didn't want the words to fly apart too fast. To this, voiceover narration and music were added.

Let's look at the movement created and the reason for my madness. The background (layer one) moved diagonally from right to left. The product (layer two) expanded and moved from top to bottom. The text (layer three) crawled and expanded horizontally from left to right. It wasn't a hodgepodge since each layer of movement complimented another layer. This was important because you don't want to have things flying all over the place with no rhyme or reason. You are only limited by your imagination and almost never by your equipment.

#288 How to Edit Music and Sound Effects

Music and sound effects are part of the editing process called sweetening. You are attempting to fine-tune your video with small nuances to make it sound superior. Like adding sugar to coffee, this sweetening makes the video "taste" that much better. Effectively using music and adding sound effects are two ways this can be accomplished.

Music chords or stingers at the right moment (a few frames before the start of the edit) will make your audience jump, grab their hearts, and scream. Soft flowing music under a love scene makes viewers more relaxed and absorbed in what they are watching. Hard-driving music during a chase scene gets the adrenaline flowing in all who are watching. Music sets the mood of any video and using the wrong music will definitely change the atmosphere.

Sound effects are much the same way. Through a process called Foley, sound effects are enhanced (and raised in volume) so the viewer is more aware of them. As mentioned earlier, a Foley artist watches the action onscreen and duplicates the sound effects right there. From walking, running, hitting, fighting, and sleeping, every sound is slightly amplified and made that much clearer. This process is important in six-channel surround sound.

Long before Foley was developed, Sergio Leone produced many spaghetti westerns (*Once Upon a Time in the West*; *The Good, The Bad, The Ugly*; and others). Because most of the soundtrack was recorded in Italian, very little dialogue was spoken, so the dubbing process to English would be that much easier.

Instead, the volume of the sound effects was greatly enhanced. The ticking of a clock, the rattle of spurs, the cocking of a gun, and footsteps in the dirt were almost deafening. They really called attention to themselves and these sound effects are what people remembered about these pictures: Almost every sound was larger than life.

You don't have to go to these extremes in your video. Do a little bit of your own Foley and rerecord an effect that is too quiet. Without calling attention to itself, it will enhance the sound of your project.

#289 Options for Deliverables

A deliverable is a word that every video producer loves to hear. This is something that is given (or delivered) to the client or end user. Final payment is usually made when one provides a "deliverable." Let's discuss the different types of deliverables.

A deliverable for a Hollywood film is currently either a VHS or DVD copy. In theatres, the 35mm release print is the deliverable that is shown in front of the audience. Former theatrical options for Hollywood and other independent films were the defunct 16mm release print for schools, organizations, and military use; Super 8 release prints for home viewing; laserdiscs (videodiscs) where the image was recorded on a 12-inch laminated plastic disc (like a DVD on steroids); RCA Selectavision where a video "record" was inserted into a player and a stylus needle began playing the picture and sound; and, lastly, Betamax, Sony's version of the home theater tape machine.

A deliverable for a video producer like you would be a copy of the completed program for the end user to watch. This could be videotape like VHS, digital video (of some sort), Betacam SP, or Digital Betacam, or it could be a disc version such as DVD (DVD-R), CD-ROM (CD-R), or CD Video. It could also be an Internet version, such as streaming video. Each of these methods enables the end user to play your video. Obviously, new formats will be added in the future, but they still will be called deliverables.

Chapter 35

Character-Generated Graphics

Solutions in this Chapter:

#290 How to Create Titles

Every video has to have words on the screen at some point. Whether this is a title or even credits, a correct way exists for displaying titles, and one must know how long to keep them onscreen.

When a title is onscreen explaining or describing something, it should be written in clear, easy-to-read text. Solution #291 will go into more detail on font choice, but if the words are difficult to read on a busy, too-dark, or too-light background, it helps no one. As an example, Kodak chose the colors yellow and black for their film boxes because it is the easiest on the eye and stands out in a crowd. It doesn't really matter if it is yellow words on a black background or black letters on yellow. The rule of thumb is the background should be dark if you want to call more attention to the words. Black is extremely dark so you can easily focus on just the yellow letters. If the background is more important and the text less, then go the opposite way.

Before the advent of color, silent films always used a black background with white text (they used black and white film—what could they do?). This was relaxing to the eye, contained no frills, and the audience just read the words. Sometimes the edges had a little scrolling, but they were always white.

A very loud color background like red is video's worst enemy. If you want to know which color to avoid in video, red would be it because it tends to bleed (no pun intended). Horror films often use black backgrounds with red letters (to simulate blood) because reds in video smear. This smearing looks like blood, and everybody's happy.

The text should appear onscreen long enough to be read twice by the viewer. Read the words to yourself slowly two times and then move on to something else. That is the established amount of time for someone to comprehend what is written. Some will read faster, others slower, but two times is the average.

Don't try to squeeze too many words onscreen or you will scare people. If necessary, use two or more screens with text. If one screen dissolves into the next, people will keep reading until you want them to stop. The flow of the eye is left to right, so text in vertical columns will be harder to read (and take longer).

What about justification? Should words be left, right, or center justified? Unless printed in a book, words on a screen should be left justified only. The right margin should not be justified. Center words onscreen if it is the title of your video or credits. People don't like reading things that are centered because it takes longer. The eye starts at the left side and has to scan right until it finds a word. It continues to do so every line in a centered message and wastes valuable screen time. Why do you think end title credits are centered and move so fast? They have too many names and don't want you to dwell on any one name too long. This why the Evelyn Wood Speed Reading course works so well; the eye is moving faster.

Do you capitalize every first letter of a word, have them all in caps, or all lowercase? Unless it's a title of something (a video, book, and so on), everything should be in lowercase except the first word of a sentence. Basically, the standard rules of English apply in reading credits and great books.

Also don't use roman numerals. We live in the twenty-first century and they don't apply much anymore. The only time to use them is in the copyright date on a video. It's done this way so no one can read them. By the time they figure out what MCM means, the date is gone. Just keep all your text in the "title safe" area and you will be fine.

Every *nonlinear editing* (NLE) system has a titling package, some being better than others. Look for a software program with some substance that enables different kerning, spacing, justifying, positioning, and color options.

#291 How to Choose the Best Font

The title of this solution should really be "Don't Use Any Font Whose Name You Cannot Pronounce." Literally millions of font choices are out there and most of the 999,986 should not be used in video.

The reason for this is that the character's edges, lines, and little scrolling may be too hard to see on the video screen. Extremely fancy fonts push video's resolution to the limit, something feature films shot on film don't have a problem with. In video, keep the font plain and simple, and it will get the point across faster, be easier on the eye, and be less frustrating and jarring for the viewer.

Block letters with a drop shadow or a slight edging work fine. They have been used forever because they work. Helvetica, Times Roman, and Ariel are block letters that fit well on screens. Choose a font size that's big enough to read, usually not more than 72 points. Tiny font sizes should also be avoided because at a distance they are unreadable. A font size of 8 or lower is only used when car commercials are discussing lease particulars and, once again, they are small because they don't want you to read the fine print. It has to be there by law, but it doesn't have to be easy to read.

Bolder fonts work better than thinner fonts because a video screen is made up of pixels. The more pixels, the sharper the picture. The more pixels in a font . . . I think you get the point.

Use underlined when you want to call attention to something and use *italics* when the words are *unique* or *special*. I use the **bold** feature on every single screen character I type. It just makes it that much thicker so it has more pixels to back it up, and you know what that means.

Just remember, if it doesn't look good to the eye and isn't easily readable on your computer screen when generating it, it won't work in the video.

#292 How to Do Credits Effectively

Credits are sometimes the only thanks people get when working on your video. Proofread everything and make sure you spell their names correctly. Great credits are remembered because they are neat, easy to read, and a source of pertinent information.

The person's first and last name should be first letter capped. Sometimes the entire name is spelled in capital letters, with the position following it, such as the title "director," being in all lowercase. This calls attention to the person's name, not their title, the way it should be.

As mentioned earlier, this is one time when centering the text is acceptable. Credits can be done in hundreds of ways and they should be done to suit your particular video. Scrolling text, moving from the bottom of the screen to the top, is the most widely used, with others fading the particular title in and out. They can also slide on from the left or right, zoom in or out, or be constantly in motion when onscreen.

Add a little life to your titles by having them grow larger or smaller onscreen, but have the movement be almost unperceivable. The movement is so slight or slow it appears to be stationary. The movie *Seven* did a great job with unusual titles that shook, blurred, vibrated, and were difficult to read, almost as if designed by a madman (which was the intent).

Design your credits to suit your video. Have fun with them by using animation, one title overlapping another, or have moving video underneath. There is no right or wrong way; just spell everything right.

Chapter 36

Business Obstacles

Solutions in this Chapter:

#293 How to Overcome Business Obstacles

Like anything else, video is a business that is changing almost daily. If you fall behind the changing trends, you no longer have an advantage. Upon graduation from college, I was trained in what motion picture film could do. My last semester in grad school, I was introduced to a newcomer: videotape. There's nothing like being trained in the idiosyncrasies of film production only to learn that video has changed the way one shoots and edits.

In this chapter, we'll discuss how to rent equipment that is state of the art without having to purchase it, how to promote your talents and services, how to maintain and update a demo reel and resume, how to look for and hire crews and talent, when to use an accountant, and how to succeed in this new medium of digital video.

#294 How to Rent Expensive Equipment

A time will come when you need to rent better equipment than you currently own. But if your budget is limited, how can you afford to rent film, digital video, or high-definition equipment without breaking the bank? The key to this is timing.

Rental houses charge by the day and the week when loaning equipment. If you rent something at $1,000 a day, the weekly rental cost (for seven days of use) will be $4,000. Most facilities charge a four-day rental week. This will help if you need a camera for up to seven days. But what happens if you are shooting a TV spot and only need the camera for two days? You could spend $2,000, but let's look at a way around that.

When renting equipment locally or out of town, you need to check it out before you begin using it; use that to your advantage. The rental house will want you to be familiar with the equipment and will get it to you before your shoot begins. For instance, if you schedule your shoot to begin Saturday morning, rent the camera package as if you need it for Friday afternoon. Pick up or have the camera shipped so you receive it Friday morning (it needs to be worked with before Saturday's shoot). You could then actually begin shooting Friday afternoon. If you work an eight-hour day, you can stop by 9:00 P.M. You now have all day Saturday and Sunday to use the camera (paying only for Saturday's usage) and return it on Monday.

This isn't sneaky and is perfectly legal because the rental house is closed on Sunday. As long as the camera is returned at the assigned time on Monday, you have finished your two-day shoot, had the camera four days, and only paid for one day's usage.

One a recent shoot, I needed the camera on a Monday. As it worked out, the particular Monday was after the Thanksgiving holiday. Because everyone is closed Thursday through Sunday, I needed to collect the equipment on Wednesday. In fact, the rental office closed at noon, so I had the camera from Wednesday afternoon and dropped it off Tuesday morning (after the shoot) the following week. This is the only time of the year where holidays make renting inexpensive. I had the camera package five and a half days and only paid for one.

Another way to save some money on camera rentals is to get two or more clients to share the rental costs. If you shoot two spots in high-definition back to back, split the cost of the camera rental between both. Since you've rented the equipment for Martha's House of Wax, why not use it to shoot Uncle Jethro's Car and Truck Emporium commercial? By scheduling both over the weekend, you still only need to rent the equipment for one day. That same one-day rental now costs even less.

If your budget is miniscule, tell the rental house. They may be able to lower your rental fee in order to get the business.

#295 How to Promote Your Services

You must constantly be selling yourself in this business. You know you have the talent; it's just a little tougher letting the rest of the world know about it.

My philosophy has always been that the best way to learn (and get better) is by doing. Books help (like this one), but real world experience is still the best teacher. My advice in promoting yourself is just go out and do it. Don't let anyone tell you, "You don't have the experience, knowledge, or funds." By doing it, you will get the experience and find out what works and what doesn't. Digital tape is inexpensive. If what you shot is junk, erase it and try again. But don't you be the one that makes that decision. Your junk is someone else's treasure. Keep an open mind and get opinions from others (opinions are free; that's why everyone has one).

If you can't sell yourself and your abilities, your digital videos won't have a chance of making money. I've produced a lot of garbage (it's collected every Wednesday), and I've learned along the way. Know your strengths and weakness, but only promote the strong points while secretly working on the weak.

Take what you have and work with it while selling yourself. "Look at the camera work in this scene." "None of these actors had ever acted before I got a hold of them." Don't be egocentric and say how great you are; just point out your strengths.

In order to best promote yourself and your work, you need to have something to show someone. You may be the greatest camera operator around, but you must prove it. The demo reel, which we'll discuss in more detail in Solution #297, is one of the best tools in displaying your various talents.

You first need to get your name in front of people and then you can show them what you can do. How do you get your name out there? Several ways are possible: knowing people, word of mouth, advertising, referrals, and business cards.

Let's look at an approach for promoting yourself as a camera operator. The same steps apply for any position on the crew. Knowing people is the easiest way to promote yourself. Once people have seen you with the camera and know you two are inseparable, the next step is to remind them exactly what you do. Tell everyone you know (and some you don't) that you make videos. Telling people is a great icebreaker and most will say, "Oh, really?" or "That's interesting," or other trivial phrases unless they have an immediate need for your services.

This word-of-mouth approach gets your name out in the business. Your objective is to get contacts, but you really have no idea who a contact may be. In actuality, anyone you meet or speak to for any reason is a potential contact. More than three-fourths of my freelance video work come from contacts I don't even know I've made. I've told someone in passing what I do for a living while seated on a plane, someone has seen me shooting something in a public place, or I've been referred to them by a friend (I only have two friends, but they work really hard).

The best way to promote anything (yourself included) is to let the world know exactly what you want them to know: advertise. Since everyone is media savvy these days, we are aware how advertising works. You need the same approach; leave a lasting impression but also promote yourself so that others will remember you.

Advertising takes on many forms: written, aural, and visual. The easiest and most widely used form of advertising is the written word or ad. The phone book with its Yellow Pages lists businesses in a single line or in various sized ads. Most will call the larger advertisement because it catches their eye, the company is probably larger and knows more (because they spent more money for the big ad), and more information about the company is listed without you having to ask them numerous questions.

The only way to get referrals from contacts is to have pleased someone with your past performance. If your last freelance employer or client liked the job you did for them, they will, without hesitation, refer you to another colleague.

A tremendous video referral service and a way of promoting yourself is your state's film or video commission. Every state in the union has a film or video commission whose sole job is to locate and refer people, crews, and services to those seeking just that. When someone comes to Pennsylvania from either in or out of state, unless they are in the video business, they don't know who to contact. Your state's video commission handbook will list your name and pertinent information (usually free) as well as others throughout the state.

Business cards are an old-fashioned but still effective means of letting people know about your talents. You can also cram a lot of important information as well as artwork on that little piece of cardboard.

Business card video CDs have established themselves as the only card that, once used, will not be discarded. This is more than just a novelty; this is a great advertising tool. This novelty is an effective tool to promote yourself and show your demo reel in the process.

#296 How to Maintain and Update Client Lists

I've always wanted a little black book and now I finally have one. It's filled with names and numbers of important people I call when I'm looking for a good time—making money in digital video.

Whether this list is on computer, notebook, Rolodex, or shards of paper, you will need to refer to it often, so be organized. Make each entry detailed enough so you know what the person does. Include pertinent contact information such as the person's business phone, cell number, fax, pager, and e-mail address. In this day and age, if you want to get a hold of someone, it's easy to do. We are never truly unobtainable (insert maniacal laughter here).

After every video project you do, write down your contact's information. After over 20 years of doing this, I have compiled quite a list. Some jobs have been small and lasted only a few hours; others have been epics. Each name should be listed in your files, no matter what the size of the project.

Contacts can get projects of various sizes, so don't assume you will always get picayune work for a certain client.

This list of your work not only helps you remember contact information, but it will become a resume of your video experience. I break mine down by year, but you can organize it by job (director or cameraperson), location (Ford Motors or Sally's Dress Shop), client's name (Uncle Vinny or Dr. Love), and others. The key thing is to be able to access it and find what you're looking for at a moment's notice.

This is a list that needs to be updated constantly. Every time you have a shoot, add that contact information to your list. Even if you never want to work for them again, you should note the information so if called again, you remember that they never paid you, they set your camera on fire, or they worked you 23 hours a day. Whether good or bad, keep track of every client or contact you meet.

Now that you have a new listing of contacts, talk to them occasionally to keep the list active. It's embarrassing when I call a contact with potential work after a four-year span only to find they moved two years earlier. This is now a dead contact. Even if you just send an e-mail every few months, let people know what you are doing, if you are available for work, and ask what they are doing. Everyone loves being asked what they are currently doing and maybe you could fit into their plans. It never hurts to talk to your contacts; once you stop, they may cease being contacts. You know what happens when you lose a contact—you can't see!

#297 How to Maintain and Update Your Reel and Resume

Like I mentioned in the previous solution, constantly update your resume by adding your job experience immediately after working for that contact. This way your resume will always remain current.

The format for writing a resume isn't as important as the material in it. My resume is three pages and lists my education and job experience. Since I have also freelanced over the last 20 years, listing every job would make my resume cumbersome. I've developed a resume that lists everything I've ever done in video, from writing to marketing. The resume is 21 pages long, but never send something that length without asking. No one wants to wade through more than three pages of text, so keep the one you send to that length. If they request more detailed information, send them the longer version.

The demo reel is no less important than a resume because it shows examples of your work. This reel should be no more than five minutes long, even if you direct feature films. If someone requests to see the entire picture, send it, but only if they ask for it. It's also good practice to get a copy of every project you are involved with; this is an example of your work.

Demo reels can be done in various formats, but one that has worked for me is structured as follows: Fade up to a title screen that contains your name, address, phone number(s), and e-mail address. This should be the first thing the viewers see so they know whose reel this is. Then fade into a title of what they are about to see and what your role is (Director: Jane Smith). It could be a reel of just your editing, directing, or shooting experience, but you need to provide the title of the program as well as what you did on that project. You can also have a different reel for every skill if you're a freelancer, but keep each one to five minutes or less. The goal is to have them wanting more.

If your reel is commercials, you can show the entire spot, but if you've done a lot of them, it might be better just to show excerpts. Constantly update this reel as you do newer and better work. I've done over 800 commercials and my reel has changed almost 50 times. Editing is easy on an *nonlinear editing* (NLE) system so update it often.

Fade out of every excerpt, fade into the title of the next piece, and so on. Never cut immediately to another piece without giving some kind of identification. The viewer needs to know if they are watching something else. Here are some other nevers: Don't just use a tape deck and pause between material samples; this leaves snow and is unprofessional. Don't use someone else's work if you didn't work on the project; that's lying and it will catch up to you. Never show too much. Anything longer than a two-minute segment will bore the viewer.

After you complete the demo, leave the viewers with your contact information once again. The goal is to hammer this vital info into their brains so they have to hire you. Your name should be the first and last thing they see.

Lastly, present your demo in a professional manner that includes labeling the tape and case with your contact info; labels that are handwritten look amateurish. Try using color or a digital photograph on your labels. Check for typos on the tape as well as in the demo reel.

Present your demo on DVD if possible. This adds an air of distinction to your work. If your demo stands out from the others they receive because of the packaging, you have a better chance of having it seen. Include business cards and follow up with a phone call after a few weeks.

It's unfortunate that just selling yourself isn't enough. The selling will get you in front of the right people, but the reel will get you the job.

As your skills increase, update your reel to reflect your improvement. Even getting that first job demands a reel. The catch-22 of getting a job without a reel if you need a reel to get a job is easily overcome. Go out a shoot a demo reel. People are more concerned about what you can do rather than who you did it for. Even if you are the client, just shoot a car commercial on your own, or anything, a compellation of your best work will get you noticed. This is a great place to use all the video footage you shot honing your skills. Put it together in a five-minute or less package to highlight what you do best. You now have a demo reel. This is your calling card, so make it count.

#298 How to Hire Crews and Talent

Hiring talent and crew demands a lot of attention. You obviously want someone who can do the job unaided without you having to watch over their shoulder. How do you find someone you can trust?

When I was in college, if I needed a female over six feet tall with red hair, I could probably find her on campus. It's actually easier today because the Internet will display a photo with every name entered. Through databases, you can easily search until you find the correct match. Acting database. Type in "actors" on a search engine and check out the listings.

When I hire people for a crew, I look at their reel, but I also ask them to shoot something for me. I learn two things from this experience: one, if they can really do what they say or their reel says they can do, and two, it allows me to see how they react in a spontaneous situation. Although shoots are usually well-planned affairs, sometimes you are thrown a curve ball and you must react quickly. By putting them in the same scenario, I see how they respond under pressure. I've had people pass the first stage but fail the second. Of course, nobody is perfect, but you get a much better idea of the person though this little test.

This is the same principle when auditioning talent. You give them a script and see how they perform reading for the role. By putting your crew through the same paces, you notice how they will perform. If someone has worked for you in the past and you are aware of their abilities, this doesn't apply.

Finding talent is easy through newspaper advertisements (crew calls), casting agencies, theater groups, or colleges. Have all talent audition and if someone doesn't work out for a current role, keep their information for a future video. Crew members can be found through local contacts, film and video commissions, or the Internet. Once again, screen all applicants to better determine their abilities. The Internet has local as well as national crew listings, but don't let popup ads for crew sway you. Anyone can shell out money and create an ad, but a reel better shows their ability.

If a potential crewperson doesn't have a web site, it doesn't mean that he or she isn't any good. He or she may be in the process of creating or updating the site, or some people just aren't computer savvy. However, if the listing leads you to a disconnected number, a web site or address is the only recourse in contacting them. Publications will become out of date quickly (like a phone book), whereas e-mail is usually more constant.

After a while, your cast and crew become like an extended family. You learn each other's likes and dislikes and will soon learn what each person excels at.

#299 When to Use an Accountant

The more you freelance in this business, the stronger your need will be for an accountant. When I worked just one steady corporate video job in the old days, my taxes were simple. But now I have over 50 different employers sending me 1099 forms. This is an accounting nightmare.

Software packages like Peachtree Accounting, Quicken, and Money enable you to input income and expenses into categories for any occupation. I've relied on Quicken for years and just print out my itemized statements.

If employed as a freelancer, you must file taxes quarterly. If you pay the government too little, you are fined, and if you pay them too much, you have little money left over. Someone has to file these quarterly payments, either you or someone else.

With tax preparation packages like Turbo Tax, you can import your income and expenses directly from another software program. This is what I do to save a few dollars, but I still have an accountant check everything over before I submit it. This is a good practice because an accountant knows more about the tax laws than I do (or most programs for that matter). It is cheap insurance to have an accountant perform at least this final step.

If you feel uncomfortable doing any of the tax preparation, have the accountant handle all of it. It will cost more money, but all you have to supply is invoices, 1099s, and receipts. It's up to you, but in this business, unless you are willing to take the time to do all the paperwork yourself, have an accountant at least look over what you've done. If it doesn't work out, maybe you can share a cell together in the big house.

#300 How to Succeed in Digital Video

This solution sounds like the title of a Broadway production (but it would also say *Without Really Trying*), but you are the only one to make this show happen and you should try. What does it take to succeed in digital video? The first easy answer is knowing and somewhat understanding the medium. Once you know what it can and can't do, you are most of the way there.

Read and learn all you can about this subject. One of the best ways I've learned is by working in the medium. As I mentioned before, my training was in film production. Although similarities exist, many more differences are apparent. I had to learn as the industry evolved.

Another great learning tool is renting DVDs of completed movies. Watching the film itself may give you inspiration, but the added bonus features are the best teaching tool. The accompanying director, writer, or actor commentary will tell you why something is happening. You learn by watching and doing, but understanding why something is done is even more helpful. Watch every extra on a DVD.

As an example, I rented *Frailty* on DVD. The movie was well written and directed, but I learned two new "solutions" in the Special Features section. The first was called the "Mag light trick." (If I had added these two solutions with correct credit noted, the book would have been called *305 Digital Film Making Solutions* and some might have thought it was a sequel.)

The Mag light trick involves putting two Mag lights on a stick and having someone dressed in black walk around your car set simulating the headlights of another car. You've all seen the effect before; two dots of light (simulating headlights) move up behind your vehicle and move off to the side. Since the car you're shooting in isn't moving and the scene occurs at night, these two Mag lights on a stick actually look like headlights approaching from behind. Maybe I'm dense, but I never really thought about that. It's a cheap, no-budget solution to a common video effect.

The second was rain effects on the side window of a car. By spraying water against a window, you get the droplets (I figured that out before I viewed the DVD). Bill Paxton, the director and lead actor in this film, wanted the droplets to move rearward as the moving vehicle (actually stationary) traveled in the rain.

When a high-velocity fan was pointed at rain-laden window, the droplets curved downward, not the way the storyboard stated (it's nice that professionals still follow storyboards). Bill wanted the rain to move sideways, like streaking rain does on a *moving* car. He ended up using an air cannon (hair dryer) to move the droplets in the correct pattern. I would have thought a fan would work too, but droplets curving down aren't realistic.

My point is by learning new things, you will succeed in this business. You already have the desire, the talent you may already possess, and the last card missing is the knowledge. Try to get as much of this knowledge as you can from others because most people are already too full of it—knowledge, that is.

Chapter 37

Comparisons

Solutions in this Chapter:

#301 How Film Compares with Video

The two battling formats, film and video, have coexisted for many years. Video has made great strides and has changed drastically from its inception. Film, on the other hand, has changed very little in the last 80 years with film speed and grain being the biggest improvements. Can you really compare the two?

Besides the obvious difference that film is developed and printed while video is immediate, let's look at the subtle things that separate the two. The biggest is *quality*. No one will argue that film is sharper and clearer than video. The best *National Television Systems Committee* (NTSC) video signal in the United States is 525 lines of horizontal resolution. The more lines you have, the sharper the picture. *Phase Alternate Line* (PAL), Europe's standard, is sharper at 625 lines of horizontal resolution. High definition, currently the best and sharpest video has to offer, is 1,080 lines of horizontal resolution, a marked improvement.

Sixteen-millimeter film, still used by documentary filmmakers, is about 2,500 lines, although film isn't really measured in lines of horizontal resolution. Film is more concerned about the size of the grain (almost like the number of pixels in a *charge-coupled device* [CCD]) in a frame of film. The smaller the size of the grain particles, the sharper and clearer the image. Since resolution is more standard, I'll continue comparing film versus video that way.

Thirty-five-millimeter film, the standard used in TV and the motion picture industry, is over 5,000 lines of horizontal resolution. Precise detail can be seen anywhere in the picture. This is still almost 10 times the resolution of NTSC and almost 5 times that of high definition.

Seventy-millimeter film, now used infrequently, was formerly the big-budget theatrical presentation format. With over 10,000 lines of horizontal resolution, it was unmatched in quality (although extremely expensive). Only IMAX is sharper because it is the largest motion picture stock available. I won't even discuss the resolution on this beast.

The other main difference between the two competitors (besides the obvious I mentioned earlier) is the contrast range. Contrast means the differentiation in luminance (brightness) between the darkest blacks and the brightest whites. In video, you have a four-stop range (4:1), meaning that the darkest objects can only be four stops less than the brightest. If a face is exposed as F4, then anything brighter than F11 will start to "clip." You've seen this in video when the camera tries to compress any bright object. Video cannot overexpose; it just compresses or "clips" the image at 100 IRE. Blacks are normally 7.5 IRE on the same scale. Basically, the picture just doesn't look right.

If you open the iris too much, the blacks become muddy and pale, showing little detail. Scenes shot at the beach are difficult at best for video because of the contrast range.

Film has an eight-stop contrast range (8:1) between blacks and whites. You can still see detail in the blacks throughout the range before they become muddy. Crisp blacks and sharply detailed whites are what make film superior in this area. However, the human eye has a contrast range of 72:1, which obviously is the best. Nothing comes close.

So which is better, film or video? You can't really compare the two that way because film is obviously sharper and has more latitude than video. Video is cheaper, more immediate, and easier to use. So it depends on the situation. I've used film for 25 years and I prefer it. But I use video in 98 percent of my shoots because it works faster, it's less expensive, and people need the immediacy. Both will be around and live happily for quite some time.

#302 How Analog Compares with Digital

The big question on everyone's mind is, "What's the difference between analog and digital?" When it comes to videotape, not much difference exists, except in the running time of the cassette, the picture quality, how it's recorded, and the lack of dropouts. For how we use it, no difference exists. Analog and digital video are shot the same way. The separation begins in audio tape.

When recording sound for video or film, care and attention has to be placed on the little things as well as the larger ones. It doesn't matter if you're working on a low-budget video or a Hollywood feature, pay attention to your sound and you will be loved by millions.

I recently had the opportunity to work on a feature film with a Toronto crew. The crew was shooting on location and they needed camera as well as sound crews. Although this was a $20 million-plus budget, most of the money had already been spent before the Toronto team arrived in Pennsylvania. They still needed top-notch people and equipment, but had very little money to spend on the local segments.

I had the luxury of being assistant sound (I prefer that rather than being called a sound ass). Almost every feature film is recorded with *digital audio tape* (DAT). Analog Nagras are seldom used, but they are still the best in the business.

The DAT machine we used was a Tascam DAP-1 field recorder and our microphone was a Sennheiser 416 shotgun, which has an extremely narrow pickup pattern and is used often in location sound. This is a basic sound package, but the "feature people" wanted perfection. I'll give you some tricks that helped me survive with the feature people.

Our first scene was recorded in a train station. A 1909 steam locomotive was to pull out of the station and fly past the camera. The engineer was hanging out of the window and the camera would be following him. Our job on sound was to follow the train so it would appear as if it roared past from your left ear to the right. Being middle-aged and enthusiastic, I thought this would be easy. I was wrong.

With the amount of noise a steam locomotive creates, sensitive shotgun microphones have to be protected (and so do their operators). Always encase a shotgun microphone in a windscreen or zeppelin. Although aurally transparent, these devices protect the delicate elements of the microphone from being damaged by wind or shocks. If you shock a microphone by telling it "You're ugly," you won't shock it if it's in a shock mount and zeppelin.

Steam locomotives also burn coal, so as it moves down the tracks, bits of unburned coal and soot are sprayed into the atmosphere. This black shower covered the zeppelin, but the 416 inside remained clean.

Another reason for using wind protection outside is the purpose it was intended for: protection from gusts of wind. Trains always make a lot of wind noise. Without a windscreen, this roar is indiscernible from any other type of loud noise. The sound operator would have to slowly raise the gain as the train approaches (keeping an eye on the level and not allowing it to distort) and then allow the sound to fade naturally without turning it down.

One area where DAT or digital audio differs from analog is in the recording level. In school (back when analog was all that existed), you were always taught to keep the level near zero on the view meter. "Loud sound means better sound," I was always told. The opposite is true in digital. You shouldn't let the sound level get to zero on the meter. If it does, it may distort or compress. Instead the best level is between -12 and -6. Maybe that's the reason Tascam's meter ends at zero.

Another thing I should probably discuss here is presence (not what you get for Christmas). Some say that digital has a colder, harsher, less inviting presence than analog. Playing a record sounds warmer and more natural than a CD (except for the clicks, pops, flutter, and scratches). I'm not good enough to notice any immediate difference between analog and digital. I can't listen to something and say, "Oh, that's digital." Maybe someday when I'm older, but for now I'm too naive.

#303 How High Definition Compares with NTSC

I've saved the best and most controversial solution for the end. *High-definition TV* (HDTV) is still the new kid on the block and is constantly growing. NTSC has been around since 1939 and the only major improvement has been color (sound, broadcasting, digital, and other factors not included).

At this writing, HDTV currently has three different standards: 24p, 720p, and 1080i. All three are immense improvements over NTSC. The first standard, 24p, is 24 frames a second progressive (as opposed to 30 frames a second in NTSC), meaning the images are progressively scanned on the screen. Digital camcorders offer this feature in addition to NTSC. A progressive scan is a building process where one line appears onscreen and then another (it happens in a millisecond). These work together to create the frames that occur 24 times a second.

This is the format in which George Lucas shot *Star Wars Episode Two: Attack of the Clones*. Very close to 35mm (also 24 *frames per second* [fps]), this format can easily make the conversion to 35mm film for theatrical release. Before the advent of 24p, a process called 3:2 pulldown would transfer the film image at 24 fps to video's 30 fps. This meant duplicating a few frames to achieve the 30 from 24. Fast pans never look good because frames are added, taking away from the smooth, flowing feel.

24p looks like film because it runs at the same speed and the viewer associates the slower frame rate with the film look. This doesn't make your videos look any better, just more like film.

The benefits of 24p HDTV are that it costs much less than film, you immediately see the results, and digital manipulation is easier because you are already in a digital medium. With film, you still create the digital improvements (characters, sets, and special effects), but that somehow has to be incorporated into the film (another step losing quality, time, and money in the process). 24p is the most widely used format when transferring video to film or for large group viewing. In addition, all HDTV images are 16:9 (16 units wide, 9 high) which is much closer to widescreen (letterbox), rather than NTSC's aspect ratio of 4:3 (4 units wide, 3 high). The public is slow to accept any change, and this is a big one.

720p is 720 lines of resolution also progressively scanned. This video image still looks great, but if transferred to film for theatrical release, it won't have the look of film that people expect in video. They believe it still looks too sharp or live for video. This format runs at 30 fps.

1080i is 1080 lines of horizontal resolution interlaced, like the fingers of two hands weaving together, where one "field" or "odd" field builds on an "even" field. It takes two of these lines, and odd and an even, to make one line of interlaced resolution. Therefore, a video frame is made up of odd and even fields (no relation to Sally Fields). The number of lines is greater, so the picture quality is sharper.

A great many other differences exist between these "big three" standards of 24p, 720p, and 1080i, but that could fill a book in itself.

You've seen HDTV broadcasts and the image is vastly superior to NTSC. By 2007, all television stations are supposed to be broadcasting in HDTV, with those who don't have HDTV sets needing to purchase a converter to view the signal on an older NTSC set. DVDs are already compatible with both standards.

This is a big deal for TV stations. Every camera, recorder, player, and TV monitor must be replaced with HDTV units. That includes all tape stock and transmission equipment. We are already broadcasting in digital, but digital HDTV involves so much more.

In the next few years as we make the transition into HDTV, watch prices come down on NTSC digital sets. Now is the best time to purchase this NTSC equipment because once the other contender hits the streets, it's all but over for the incumbent.

Index

P

Q

R

S

T

About the Author

Since 1980, Chuck Gloman has been an independent producer, videographer, director, and editor with experience in all areas of video—from commercial production (800 to date) to corporate training (450 to date). A resident of York, Pennsylvania, he is a regular contributor to *Videography*, *Mix*, *Television Broadcast*, and *Government Video*.